Fundamentals of
Behavior Analytic
Research

APPLIED CLINICAL PSYCHOLOGY

Series Editors:
Alan S. Bellack
University of Maryland at Baltimore, Baltimore, Maryland
Michel Hersen
Nova Southeastern University, Fort Lauderdale, Florida

A Continuation Order Plan is available for this series. A continuation order will bring delivery of each new volume immediately upon publication. Volumes are billed only upon actual shipment. For further information please contact the publisher.

Fundamentals of Behavior Analytic Research

Alan Poling, Laura L. Methot, and Mark G. LeSage

Western Michigan University
Kalamazoo, Michigan

Plenum Press • New York and London

OBN 133/358

Library of Congress Cataloging-in-Publication Data

On file

ISBN 0-306-45056-9

© 1995 Plenum Press, New York
A Division of Plenum Publishing Corporation
233 Spring Street, New York, N. Y. 10013

10 9 8 7 6 5 4 3 2 1

Printed in the United States of America

BF
181
.P59
1995

Preface

By the end of his long life, B. F. Skinner (1904–1990) had become one of the most influential and best known of psychologists (Gilgen, 1982; Heyduke & Fenigstein, 1984). An important feature of the approach to the study of behavior that he championed, behavior analysis, is the intensive study of individual subjects over time. This approach, which is characterized by the use of within-subject experimental designs, repeated and direct measures of behavior, and graphic analysis of data, stands in marked contrast to the research methods favored by many nonbehavioral psychologists. Skinner discussed the advantages of his approach in a number of books (e.g., Skinner, 1938, 1953, 1979), but never devoted a book to methodology.

Sidman (1960) and Johnson and Pennypack (1993b) did devote books to behavior analytic research methodology. These books are of exceptionally high quality and should be read carefully by anyone interested in behavior analysis. They are sophisticated, however, and are not easy reads for most neophyte behaviorists. Introductory-level books devoted entirely to methods of applied behavior analysis (e.g., Kazdin, 1982; Barlow & Hersen, 1984) are easier to understand, but somewhat limited in coverage. The limited coverage that most traditional texts on research methodology devote to the Skinnerian approach may be accurate (e.g., Cozby, 1993), but some badly misrepresent the approach (e.g., Christensen, 1994; Reaves, 1992). Given the texts currently available, there appears to be a need for a simple and practical, yet wide-ranging, behavioral methodology text. *Fundamentals of Behavior Analytic Research* is intended to be such a book.

Purpose of This Book

Our purpose in writing this book was to provide an introduction to the use of scientific methods in the study of the behavior of humans and nonhumans. Its primary strengths are in presenting these methods in a

v

straightforward manner and in providing a wealth of practical guidelines for beginning researches. These guidelines are intended to aid beginning researchers in avoiding significant problems and in producing research that will be acceptable as the basis for a thesis, dissertation, or journal article.

We, the authors of *Fundamentals of Behavior Analytic Research*, are two graduate students and a professor. Each of us has succeeded in some research projects and failed in others. In writing the book, we constantly asked: What should one consider at each step in designing, conducting, and interpreting a study? How can problems be averted and the likelihood of producing meaningful findings be maximized? There are no simple answers. Research projects are different, and so are researchers. Nonetheless, there appear to be some common strategies in the repertoires of most successful behavior analytic researchers.

These strategies are best passed along through direct contact between aspiring researchers and successful scientists. As best we know, the most effective way to learn how to do research is to study with a good researcher who is dedicated to research training. Such a mentor will outline good rules for conducting studies and will provide access to an environment in which research skills can be acquired through trial and error. No book can take the place of a skilled mentor. *Fundamentals of Behavior Analytic Research* is intended to augment, not replace, a mentor's lessons. It is also intended to be of some assistance to those unfortunate students who are interested in research but have no productive research supervisors available.

Organization of the Book

Fundamentals of Behavior Analytic Research comprises nine chapters, each relatively short. Chapter 1 overviews science in general and the science of behavior analysis in particular. Although there is no single scientific method, there are limits to the kinds of phenomena that can be studied from a scientific perspective, and there are some general characteristics of the scientific approach to studying anything. Chapter 2 considers the selection of general research topics, the derivation of experimental questions from general topics, and the design of experiments that address these questions. It provides a foundation for issues of design that are covered more fully in Chapters 5 and 6, which cover within-subject and between-subjects designs, respectively. Chapter 6 also summarizes some nonexperimental approaches. As a rule, for reasons explained in these two chapters, experimental methods are superior to nonexperimental methods, and within-subject designs are superior to between-subjects designs.

Chapter 3 provides guidelines for actually conducting experiments and discusses strategies for evaluating their importance. These are significant topics, in that every scientist wants to conduct studies that are important and to conduct those studies competently. Chapter 4 considers data collection techniques, which are strategies for defining and measuring behaviors of interest. Chapters 7 and 8 examine two alternative strategies for evaluating the data obtained in a study, graphic data analysis and statistical data analysis, respectively.

The culmination of the research process, sharing one's findings, is the topic of Chapter 9. This chapter provides guidelines for making conference presentations and publishing journal articles.

Throughout the book, material we deem to be especially important is highlighted in *italic* type.

Fundamentals of Behavior Analytic Research is intended for undergraduate and graduate students with little practical knowledge of research methods. It is no more than a primer and therefore does not cover comprehensively any of the topics introduced. Scientific methodology, graphic analysis, statistical analysis, behavioral assessment, research design, and any of a score of other topics have each been the focus of complete volumes. The material presented here may make those volumes easier to understand, or at least prompt students to seek them out, and we will be pleased if either of these outcomes occurs. We will be more pleased, however, and our objective will be attained, if the book helps people to conduct meaningful studies of behavior. In a world full of nonsense and woe, such studies are sorely needed.

Many people contributed to this book, although they cannot fairly be blamed for its weaknesses. Because we understand the research conducted by members of the Behavioral Pharmacology Laboratory at Western Michigan University, because it is easy to get permission to reproduce the research, and because the research is neither especially good nor especially bad, we often make reference to that research. Over the years, the members of the laboratory have included Ken Alling, Eb Blakely, Tom Byrne, Linda Chapman, Rod Clark, Jim Cleary, Dawn Delaney, Sue Goeters, Deb Grossett, Earl Hall-Johnson, Maurie Holbrook, Kevin Jackson, Kim Jarema, Cathy Karas, Kelly Kent, Kathy Krafft, Malath Makhay, Mike Monaghan, Carol Parker, Vicky Pellettiere, Mitch Picker, Steve Ragotzy, Hank Schlinger, Rob Sewell, Steve Starin, Steve Sundby, Scott Wallace, Jason Wilkenfield, and Connie Wittkopp. Thanks to all of you. Dick Malott and Jack Michael have sharpened our knowledge of behavior analysis and have worked for many years to keep the field alive and well; we thank them on both counts. We thank the students of Psychology 330, who were exposed to much of the material as it was developed and refined. We hope that it is better now.

Finally, this book would not have been completed without the help of Locke Ridge Coal, chief pheasant finder and footwarmer, to whom it is dedicated.

Contents

CHAPTER 1

Science and the Analysis of Behavior

The word *science* comes from the Latin *scientia*, meaning "knowledge," but as Sir Peter Medawar (1984, p. 3) emphasizes:

> ... no one construes "science" merely as knowledge. It is thought of rather as knowledge hard won, in which we have much more confidence than we have in opinion, hearsay and belief. The word "science" itself is used as a general name for, on the one hand, the procedures of science—adventures of thought and stratagems of inquiry that go into the advancement of learning—and on the other hand, the substantive body of knowledge that is the outcome of this complex endeavor, though this latter is no mere pile of information: Science is organized knowledge, everyone agrees, and this organization goes much deeper than the pedagogic subdivision into the conventional "-ologies," each pigeon-holed into lesser topics.

The purpose of this book is to describe the strategies of inquiry that are characteristic of a science that attempts to make sense of the behavior of humans and other animals—namely, behavior analysis. This chapter introduces science in general, then considers the characteristics of the science of behavior analysis.

Characteristics of Science

Science is a search for regularities in the natural world, a search that employs many strategies. Science is what scientists do, and scientists do many different things. Despite a commonly held belief to the contrary, there is no "scientific method" in the sense of a step-by-step recipe that

Portions of the material presented in this chapter initially appeared in Poling, Schlinger, Starin, and Blakely (1990).

scientists routinely follow. *Scientists are behaving organisms, and their actions are affected by the same kinds of variables that affect the behavior of other people.* Some hold generally discredited notions, some operate in fringe areas not widely accepted as legitimate, and a few fabricate data or otherwise cheat for their own benefit. All scientists are influenced by the specific training that they receive and the cultures, lay as well as scientific, in which they live and work. Legitimate scientists often disagree with one another, and they differ dramatically in their personal styles, interests, and ethics. Nonetheless, there are limits to the kinds of phenomena that can be investigated scientifically and to the strategies that can be employed in these investigations. In general, five characteristics typify the behavior of scientists.

1. Scientists are concerned with empirical phenomena. The natural world is the source of all scientific information (*data*), and that world consists of *empirical phenomena.* These phenomena are objects and events that can be detected through observation. Most scientists agree that good data are objective, reliable, and quantitative. Data are *objective* if the events of interest can be observed by more than one person, *subjective* if they cannot be. They are *reliable* if independent observers can agree as to whether or not the events have occurred. And they are *quantitative* if they can be scaled in physical units along one or more dimensions.

2. Scientists disclose orderly relations between classes of events and use these relations to explain natural phenomena. The essence of science is determining whether the value of one kind of environmental event, termed a *variable* (because its value can vary), is affected by the values of other kinds of environmental variables. Often, though not always, this determination involves experimentation in which the scientist manipulates the value of one variable, termed the *independent variable,* and evaluates whether doing so changes the value of another variable, termed the *dependent variable.* If the value of the dependent variable changes in orderly fashion as a function of the independent variable, the two are *functionally related.* Explanation in science begins with specifying the variables of which a particular kind of event is a function. In a lay sense, those variables are causal—they determine whether or not the event in question occurs and, if it does, its magnitude and other characteristics. When the variables that cause a particular kind of event can be specified with precision, that event is understood.

3. Scientists attempt to predict and control their subject matter. Once it is clear that an independent variable affects a dependent variable, it becomes possible to predict and control the dependent variable. Prediction is possible because the value of the independent variable determines the value of the dependent variable. Thus, all other things being equal, the probable value of the dependent variable can be predicted given knowledge of the value of the independent variable. Moreover, the value of the dependent variable can be controlled by selecting the value of the independent variable. The practical value of science rests in large part on the isolation of independent variables that are subject to manipulation and that change important parts of the world (i.e., dependent variables) in a manner that benefits humanity.

4. *Scientists assume that the phenomena they study are orderly and deterministic.* Scientific prediction is possible because some variables are lawfully related to other variables: If A has occurred, then B is likely to follow. Knowing this relation, one can predict when B will occur (after A). Such knowledge is dependent on an orderly world, one in which relations revealed in the past (e.g., B follows A) will recur. Moreover, it requires a deterministic world, one in which the values of some events depend upon (i.e., are determined by) the values of other events. These dependencies hold true regardless of the subject matter of the science.

5. *Scientists make assertions that are tentative and testable.* Observations of real events are the raw material of science. Relations among these observations form the basis of scientific laws. These laws, in turn, are related to one another through higher-order principles and concepts. Scientists recognize that observations are necessarily inexact and that important relations among variables are not easy to detect or to summarize in a meaningful manner. Therefore, scientific assertions at all levels do not reflect absolute truth, but are instead tentative and subject to revision.

It does not follow from this uncertainty that scientists know nothing or that scientific assertions are so imprecise as to be meaningless. Consider the law that describes how far a falling body will travel in a given period of time:

$$S = \frac{1}{2}gt^2$$

where S is meters traveled, g is a constant that varies with location and altitude (and thus accounts for variations in the force of gravity), and t is time in seconds. This equation enables one to predict with very good accuracy how far any object will fall in a given time at any place on earth. Other kinds of scientific knowledge allow for equally good prediction of other phenomena. In fact, when it comes to predicting natural phenomena, scientific information usually is far superior to any other kind.

In large part, this superiority obtains because scientific knowledge is not fixed, but changes over time to more accurately reflect observations. *Science is both fallible and self-correcting.* It is fallible in that assertions made at a given time may subsequently be modified or rejected; it is self-correcting because these changes lead to a progressively refined world view, one that provides better integration of observations and allows for better prediction of important phenomena. These attributes are the primary virtue of science.

The Science of Behavior Analysis

Behavior analysis is a unique approach to the study of behavior popularized by B. F. Skinner, although many other people contributed to its development. The historical development of behavior analysis has been described with admirable clarity elsewhere (Michael, 1980; O'Donnell, 1985; Zuriff, 1985) and will not be considered here.

Throughout his long and productive life, Skinner had as one of his primary goals the discovery of lawful behavioral processes that operate at the level of the individual organism (e.g., Skinner, 1956, 1966). His approach to discovering these processes had four noteworthy characteristics:

1. *Skinner eschewed the formal development and testing of theories.* That he did so does not mean that he failed to systematize findings or develop an organized set of concepts. Quite the contrary. Most of his work focused on the control of behavior by its consequences, that is, on operant conditioning, and Skinner and his colleagues succeeded in describing such basic operant relations as reinforcement, punishment, and stimulus control. What they did not do was develop formal hypothetical and deductive models intended to predict exactly how organisms would behave in certain situations.

Such models are exemplified by the complex formal theoretical system based on need reduction developed by Clark Hull (1884–1952). The system was constructed as a system of detailed postulates and corollaries, all presented in verbal and mathematical form. Most of Hull's work involved testing theoretical predictions and modifying his system to incorporate experimental results. These modifications made the system elaborate and cumbersome. There were 18 postulates and 12 corollaries in his last version (Hull, 1952), but it failed to provide a good description of the behavioral phenomena of interest. Skinner, reasonably enough, was not interested in mimicking Hull's failure.

2. *Skinner studied a few subjects intensively.* Most of Skinner's early research was conducted with rats. Later, he came to favor pigeons as subjects. With both species, his experiments characteristically used relatively few subjects, often only two or three. The behavior of these subjects was measured repeatedly over time, and in most studies each subject was exposed to every condition of interest. This approach to research involves repeated measures of behavior and within-subject experimental designs

and contrasts markedly with the nonrepeated-measures, between-subjects designs favored in psychology early in this century.

3. *Skinner determined the effects of the variables of interest by visual inspection of findings, not by statistical analysis.* Early in this century, R. A. Fisher (1925) and others developed mathematical techniques for determining the probability that differences in mean scores across groups were due to sampling error, or "chance." These techniques proved both necessary and useful for interpreting the results of large-sample, between-groups experiments, which came to dominate psychology. But they were neither necessary nor especially useful in analyzing the results of the kinds of experiments favored by Skinner, and he did not employ them.

4. *Skinner emphasized that behavior is of interest in its own right, not as a sign or symbol of anything else.* That behavior is important in itself may seem obvious, but many psychologists are primarily concerned with behavior as a reflection of some underlying process or condition, such as mental illness, or activation of a neurochemical pathway in the brain. The behavior analyst, in contrast, assumes that behavior is a rich and fascinating subject matter in its own right.

Skinner's approach to the study of behavior came to be called the *experimental analysis of behavior (EAB)*. A journal with that name was founded in 1958, and the *Journal of the Experimental Analysis of Behavior* (JEAB) continues as an influential research outlet. The book *Tactics of Scientific Research* by Murray Sidman (1960) covered the logic and practice of EAB with admirable clarity and its contents have influenced many researchers (Moore, 1990).

As one of his main points, Sidman emphasized that science is inevitably concerned with variability. Behavior changes, and the primary task of the behavioral scientist is to explain how and why it changes. The behavioral scientist does so experimentally by manipulating the value of an independent variable and measuring how some dimension of behavior, the dependent variable, changes in response to changes in the independent variable. Unfortunately, even under apparently stable conditions, behavior is rarely consistent across different subjects or within individual subjects across time, which sometimes makes it difficult to ascertain whether or not behavior actually changes as a function of the level of the independent variable.

There are basically two strategies for dealing with variability in an experiment. One strategy, not recommended by Sidman, is to accept variability as intrinsic and try to detect the effects of the independent variable despite considerable fluctuations in the level of the dependent variable even under stable conditions (i.e., when the level of the independent variable is unchanged). This search for effects is often carried out by means of statistical analysis. The second strategy, which Sidman advocated, is to isolate and control the extraneous variables that are responsible for the variability. Although he recognized that this strategy is not always possible, it helped to differentiate EAB from mainstream experimental psychology at the time *Tactics* appeared. Other distinguishing features of EAB, discussed previously in the context of Skinner's approach to research, are summarized in Table 1.1.

This table provides a convenient, albeit oversimplified, summary of how EAB differed, historically, from traditional experimental psychology. Although some of these distinctions have blurred over time, it is still the case that the majority of studies published in JEAB are EAB studies according to the criteria listed in the table. Moreover, studies with these characteristics have proven especially useful in isolating and describing fundamental principles of behavior.

In addition to emphasizing the importance of isolating and describing fundamental principles of behavior, Skinner contended (e.g., Skinner, 1953) that these principles and the general strategies of EAB could be used to benefit humanity. Following a flurry of studies demonstrating that basic operant principles generally held with humans as well as nonhumans (e.g., Bijou, 1955, 1957; Lindsley, 1956), researchers demonstrated the truth of this contention.

In one of the first studies to do so, Ayllon and Michael (1959) reported that aggression by institutionalized adult patients in a state hospital could be improved by appropriately scheduling the attention paid to patients by nurses. Subsequent studies by Ayllon and his associates demonstrated that

TABLE 1.1. A Comparison of Research Strategies Characteristic of the Experimental Analysis of Behavior and of Traditional Experimental Psychology

Dimension	Experimental analysis of behavior	Traditional experimental psychology
Number of subjects	Few	Many
Research design	Within-subject	Between-subjects
Data collection	Direct, repeated measures of behavior	Various methods, often indirect and nonrepeated measures of behavior
Data analysis	Graphic	Statistical
Approach to variable data	Consider the variability as imposed; isolate and control the responsible extraneous variables.	Consider the variability as intrinsic; use statistics to detect effects of the independent variable despite the variability.

social attention could be used as a reinforcer in the treatment of many troublesome behaviors exhibited by psychiatric inpatients (Ayllon, 1963; Ayllon & Haughton, 1964). Later, Ayllon began to utilize reinforcers other than social attention and eventually developed the token economy (Ayllon & Azrin, 1968).

Other researchers studied the use of EAB methodology to deal with a variety of behavioral problems in other settings and populations, including college students, children, parents, criminals, and people with mental retardation. Reasonable successes often were reported, as is evident in two early edited texts (Ulrich, Stachnik, & Mabry, 1966; Ullmann & Krasner, 1965). These successes, coupled with the abundance of research funds during the Kennedy era, fostered further projects in which EAB methods were used to help people.

By the late 1960s, so much such research was being conducted that a special outlet was needed for its publication. That outlet was the *Journal of Applied Behavior Analysis* (JABA), begun in 1968. This journal was specifically devoted to studies in which the characteristic strategies of EAB were applied to the solution of social problems, a type of research known generally as *applied behavior analysis (ABA)*. In the first issue of JABA, Baer, Wolf, and Risley (1968) described the then-current dimensions of applied behavior analysis.

Dimensions of Applied Behavior Analysis

The seven dimensions of ABA described by Baer et al. (1968) were primarily prescriptive. That is, they were intended, at least in part, as characteristics that ought to be present in articles published in JABA. Although the field has not been static with respect to the importance placed on each dimension (Hayes, Rincover, & Solnick, 1980; Baer, Wolf, & Risley, 1987), the seven together continue to capture the essence of the field (Baer et al., 1987). They are described briefly below. For elaboration, see Baer et al. (1968, 1987).

Analytic: An analytic study employs a convincing experimental design, that is, one that allows the researcher to state with confidence whether or not the independent variable influenced the dependent variable.

Applied: A study is applied if it attempts to improve behaviors that constitute a problem for the behaving individual or for another person with a legitimate interest in the behaving individual (e.g., a spouse or parent).

Behavioral: A study is behavioral if it measures what a subject actually does and focuses on that activity as important in its own right, not as a sign or symptom of activity at another level of analysis.

Conceptual: A study is conceptual to the extent that its procedures make sense in terms of, and are described with reference to, accepted principles of behavior.

Effective: An effective study is one in which the changes in behavior actually benefit participants. Put differently, an effective study produces clinically significant behavior change.

Technological: A technological study is described with sufficient clarity and detail to allow others to replicate it.

Generality: A study has generality to the extent that its results are demonstrated to hold across time, or across different kinds of participants, problem behaviors, or settings.

Determining whether or not these seven dimensions are adequately demonstrated in a given investigation is not easy, and equally well-trained researchers can legitimately disagree on whether or not a given article constitutes "good" ABA. Nonetheless, the seven characteristics described by Baer et al. (1968, 1987) are generally recognized as important, and aspiring applied researchers would follow good practice to consider each characteristic carefully as they design, conduct, analyze, and report research.

With the exception of the dimensions of effective and applied, which are not relevant, all of them are also important in basic research. Although the distinction is not absolute, *basic research* in behavior analysis is intended to determine fundamental principles of behavior; that is, it delineates the variables that influence the actions of living creatures. *Applied research*, in contrast, determines whether interventions based on the application of behavioral principles are effective in changing human behavior in desirable ways. For example, studies designed to explore how the scheduling of intermittent reinforcers influences rate of responding and resistance to extinction, such as those reported in *Schedules of Reinforcement* by Ferster and Skinner (1957), are basic research studies. Among other things, those studies revealed that variable-ratio (VR) schedules of reinforcement characteristically engender relatively high rates of responding and substantial resistance to extinction. Under VR schedules, every nth response (e.g., 10th under a VR 10 schedule), on average, produces the reinforcer, although specific response requirements vary from reinforcer to reinforcer (e.g., they might vary from 1 to 20 under a VR 10).

Knowing how VR schedules affect behavior, De Luca and Holborn (1992) arranged such schedules to produce substantial levels of exercising (riding stationary bicycles) by obese (and nonobese) boys. Because obesity and lack of exercise increase health risks and diminish psychological well-being, this study was applied.

Over time, applied and basic research in the behavior analytic tradition expanded into a wide range of topical areas. The contemporary breadth of the field of behavior analysis is evident in Table 1.2, which lists the 11 specialty areas represented at the 1994 Association for Behavior Analysis (ABA) Convention. This organization comprises members with a common interest in behavior analysis, although their specialties differ widely.

Members of ABA, and behavior analysts in general, are by no means a homogeneous group. Although they usually favor the general kinds of experimental strategies advocated by Skinner and by Sidman, they are by no means restricted to them. Good researchers are not dogmatic. There are situations, for example, that demand between-subjects experimental designs and indirect measures of behavior. The following chapters describe a variety of approaches to research and consider the situations under which these approaches are likely to prove useful.

Summary and Conclusions

Science provides an approach, at once unique and uniquely powerful, to gathering and organizing information about the natural world, including behavior. Although there is no single "scientific method," there are

TABLE 1.2. Specialty Areas Represented at the 1994 Association for Behavior Analysis Convention

Behavioral clinical interventions, behavioral medicine, and family interventions
Behavioral pharmacology and toxicology
Community interventions and correctional settings
Conceptual analysis
Developmental disabilities and autism
Education
Ethical, legal, and social issues
Experimental analysis of behavior
Human development and gerontology
Performance management and training
Verbal behavior

certain common features in the actions of people who behave as scientists. There are further commonalities in the actions of behavior analysts, including the research methods that they characteristically employ. These methods are useful for evaluating the behavioral effects of a wide range of independent variables, regardless of who employs them. The strategies and tactics of research advocated in this book are not for behavior analysts only.

CHAPTER 2

Designing Experiments

Many different strategies are to some extent useful for studying particular kinds of behavior and the variables that influence those behaviors. The following scheme, although somewhat oversimplified, is a convenient way of envisioning the conduct of studies of behavior as taking five steps:

1. Select a topic for study.
2. Design an experiment appropriate for studying the topic.
3. Conduct the experiment.
4. Analyze the results.
5. Present the study and its findings to an appropriate audience.

Steps 1 and 2 are the primary topics of this chapter.

Research Topics and Experimental Questions

The general topic that a researcher studies is primarily a function of the researcher's training and current position. For example, a person trained in behavioral pharmacology will in all likelihood pursue research that is concerned with the behavioral effects of drugs, the general mechanisms through which these effects are mediated, and the variables that modulate drug action (Poling, 1986). Each of these broad areas encompasses a range of subtopics amenable to experimental analysis.

A critical skill for researchers is to be able to move from general topics of interest to specific questions that can be answered through studies of manageable size and complexity. Those studies are intended to answer the *experimental question*, which Johnston and Pennypacker (1993b, p. 366) define precisely and concisely as "a brief but specific statement of what the researcher wants to learn from conducting the experiment." These authors provide extensive coverage of the characteristics of experimental questions that make them useful in behavior analytic research (Johnston & Pennypacker, 1986, 1993b).

In general, desirable open-ended experimental questions are those that delineate the specific variables of interest but do not posit a particular relation between them. Such questions help to avoid a hypothesis-testing approach to research and the limitations and problems that such an approach entails. Examples of reasonable experimental questions include these:

1. What are the effects of behavioral parent training on parent and child feeding-related behaviors in the child's natural eating environment (Werle, Murphy, & Budd, 1993)?
2. How does performance under a piece-rate pay system in which 100% of the subject's pay is incentive-based differ from performance under a base pay plus incentive system (Dickinson & Gillette, 1993)?
3. How does delayed reinforcement affect response acquisition in untrained rats (Lattal & Gleeson, 1990)?

Given that the purpose of an experiment is to answer the experimental question, one must take care to frame the question clearly and accurately. If one does not state the experimental question precisely, one cannot ascertain how, or even whether, findings relate to that question. Moreover, and equally important, one must take extreme care to ensure that the experimental procedures employed actually address (i.e., are capable of answering) the experimental question. Good experimental questions are capable of being answered, and good experimental procedures provide at least tentative answers to them.

Scientists move from the general to the specific as they reduce broad-spectrum topics of interest to tightly focused questions that can be answered through experimentation. They then move from the specific to the general as the answers to the delimited questions are used to make sense of the general topic. As an example of this strategy, we will consider a line of research conducted in our laboratory.

Nearly 15 years ago, we became aware from reading the relevant literature and direct observations of people with epilepsy that anticonvulsant medications could produce adverse behavioral effects. There were few studies in this area, however, and their results were inconclusive. Therefore, a general question of some importance was: What are the behavioral effects of anticonvulsant drugs in people who use these medications therapeutically? This broad-spectrum question eventually led to a series of studies in our laboratory (see Poling & Picker, 1987).

Gaining the information of interest required further refinement of the question. Because of the difficulty of conducting methodologically rigorous drug studies with humans, we decided to examine the behavioral effects of common anticonvulsant drugs in nonhumans, in the belief that these effects would generalize meaningfully to humans. Thus, the question became: What are the behavioral effects of particular anticonvulsant drugs as determined in laboratory studies involving nonhuman subjects?

Designing actual studies required still further refinement of the question. To conduct actual experiments, we were forced to decide on the subjects to be used, the drugs to be examined, the behaviors of interset, and the procedures to be used to examine these behaviors. This step led to the formulation of specific, and answerable, experimental questions. For example, the experimental question in one study (Picker & Poling, 1984) was this: What are the acute effects of phenobarbital, clonazepam, valproic acid, ethosuximide, and phenytoin on learning in pigeons exposed to a repeated acquisition procedure? This procedure requires subjects to learn a spatially defined sequence of responses that changes each day and is very sensitive to drug effects (Thompson, 1978).

Over the years, we conducted several studies designed to answer such specific experimental questions. By examining the collective results of these studies, we were able to profile the acute and chronic effects of several common anticonvulsant drugs on the performance of pigeons and rats responding under procedures that measured what many people would call learning, memory, reaction time, and performance. There were substantial differences in the behavioral effects of different drugs, with the older, more sedating drugs characteristically being more disruptive than the newer, less sedating ones. With repeated exposure, some tolerance (i.e., diminution in a drug's effects) usually occurred, although the degree of tolerance differed somewhat across drugs and assays. Where comparisons could be made, drug doses that disrupted behavior were not significantly larger than those required to control seizures. Summary information of this sort begins to provide an answer to this question: What are the behavioral effects of particular anticonvulsant drugs as determined in laboratory studies involving nonhuman subjects? Our results were roughly comparable to those obtained in clinical studies with epileptic humans and are relevant to, although they do not fully answer, our initial, and general, question of interest: What are the behavioral effects of anticonvulsant drugs in people who use these medications therapeutically?

In the foregoing example, published material dealing with the effects of anticonvulsant drugs provided the primary impetus for choosing a general research topic, although direct observations of medicated people also played a role. Familiarity with the literature in one's area of interest is especially important for generating useful research topics; if it serves no other purpose, it prevents the unwitting replication of studies that have already been done. Discussing possible research endeavors with established scientists is similarly valuable, and neophyte researchers will profit from discussing possible topics with successful people in the field. In fact, researchers characteristically begin their careers by studying topics provided by estalished scientists.

In general, there are four kinds of research topics:

1. *Topics based on unexamined relations.* These topics arise when there is reason to believe that two (or more) variables may be related, but it has not been determined whether they are. For example, schedule-induced defecation is known to occur in rats, and it is reasonable to speculate that the phenomenon would occur in other species. A recent study from our laboratory (Jarema, LeSage, & Poling, 1995) began to explore this possibility by answering the specific question: Does schedule-induced defecation occur in pigeons exposed to fixed-time schedules of food delivery?

2. *Topics based on practical considerations.* In applied research, the intervention of interest characteristically is known to be effective in some situations, and the experimental question involves examining whether it is also effective in dealing with another kind of behavioral problem. For example, one of our studies (Wittkopp, Rowan, & Poling, 1991) examined whether an intervention involving a combination of training and feedback would be effective in reducing machine setup time in a manufacturing plant. Our study found that it was, a finding that is consistent with the results of a large number of studies conducted under somewhat different circumstances.

Other practical research topics involve comparing the effects of different interventions and evaluating potential treatment interactions. As an example of this approach to generating research, either methylphenidate (Ritalin) or contingency management procedures alone have been shown to be useful in managing hyperactive children, and several studies comparing their effects alone and in combination have been conducted (e.g., Ross & Ross, 1982).

3. *Topics based on disagreements.* In some cases, similar studies yield contradictory results. Experiments designed to resolve such a disagreement, perhaps by controlling an extraneous variable that may be responsible for it or otherwise increasing methodological rigor, are potentially very important. For example, some studies examining the effects of stimulant drugs report different effects in people with mental retardation than in otherwise similar individuals; other studies find no difference (Gadow &

Poling, 1988). Determining what accounts for this disparity is of considerable significance, but has not yet been accomplished.

4. *Topics based on theoretical predictions.* Although much of behavior analysis is not tightly theory-driven, this generalization does not hold true in some areas, particularly those that involve the quantitative analysis of performance under concurrent schedules. Work in this area often entails developing mathematical models that precisely describe relations between behavioral outputs and environmental inputs under particular schedule arrangements, then checking to see whether the model holds under other arrangements and, if it does not, modifying the model as necessary.

As an example of this approach to generating research questions, researchers have expended considerable effort in determining the conditions under which the generalized matching equation provides a good description of behavior. For interested readers, the equation, when expressed in logarithmic coordinates, takes the form:

$$\log (B_1/B_2) = a \log (r_1/r_2) + \log c$$

where B is the measure of behavior (responses or time) under a component schedule, r is the obtained reinforcement rate under a component schedule, and the two alternative component schedules are designated by the subscripts 1 and 2. The slope of the line fitted to the data, a, is a measure of the degree to which behavior is influenced by differences in reinforcement rates (i.e., the sensitivity of behavior to reinforcement). Log c, the intercept, is an estimate of bias in behavior for one or the other schedule.

Studies of matching have played an important role in the experimental analysis of behavior, and the matching equation provides a good description of concurrent-schedule performance in many studies (de Villiers, 1977; Davison & McCarthy, 1988). Nonetheless, a potential problem with model testing is that the game can go on almost endlessly, and it is easy to lose track of whether the models developed actually are useful in describing, predicting, or controlling meaningful behaviors in humans or other organisms.

Very few aspects of an aspiring scientist's career are more important than the topics chosen for study. There is no one best strategy for selecting a general reserach topic, but the following practical suggestions can be offered:

1. *Work in areas that interest you.* The extrinsic payoffs in research usually are small and delayed; therefore, it is important that the day-to-day collection and analysis of data be rewarding to the scientist.

2. *Work in areas in which you are well trained.* Although there are certain general skills that are useful in all behavior analytic research, specialized

knowledge of the topic of interest increases the likelihood of success. Enlisting the aid of colleagues, and learning from them, is a useful strategy for extending one's scientific repertoire.

3. *Work in areas that are supported by a social community.* As discussed subsequently, the "goodness" of research is determined by a social community. Although there may be an eventual payoff for persisting in studies that initially are not sympathetically received in any quarter, the beginning researcher probably should choose somewhat conventional topics for study. Clues to such topics are available in published scientific literature, in conference presentations, and in requests for research proposals (RFPs) by granting agencies.

4. *Don't persist in folly.* Some research topics appear at the onset to be perfectly reasonable, but do not bear fruit. How persistent one should be in examining such a topic is difficult to ascertain and depends on the apparent importance of that topic, the available alternatives, and the consequences of persisting unsuccessfully. For example, a newly hired assistant professor must generate publishable data within a relatively short period in order to secure promotion and tenure. As a rule, research topics that fail to bear fruit after three or four attempted experiments are high-risk; do not devote primary resources to them.

A recommended practical strategy for beginning researchers is to conduct low-risk/low-payoff and higher-risk/higher-payoff studies simultaneously. The former are straightforward and pedestrian experiments, with good likelihood of producing results that will be deemed interesting, though not of great importance, by a social community. An example of such a study, referred to earlier, is our demonstration of schedule-induced defecation in pigeons (Jarema et al., 1995). So long as procedures similar to those used in other studies of schedule-induced defecation were used with appropriate care, results would be of interest to several scientists regardless of whether or not the phenomenon was demonstrated. Moreover, these procedures are straightforward and not difficult to arrange. Given these considerations, we knew at the planning stage that there was a good likelihood the study would "work," and it did.

Higher-risk/higher-payoff experiments characteristically involve the use either of novel procedures to investigate new topics or of procedures that are known to pose problems. A modest example of such a study is an experiment by Renfrey, Schlinger, Jakubow, and Poling (1989), in which the behavioral and anticonvulsant effects of phenobarbital and phenytoin (Dilantin) were evaluated simultaneously in surgically prepared rats. For us, the procedures were difficult, and there was considerable risk that the study would not be completed or would yield uninterpretable data. Neither of these difficulties eventuated, and the study yielded interesting and potentially important results (i.e., the doses of phenobarbital and phenytoin that disrupted rats' operant performance were in the range of the doses required to control seizures in the same animals). We were not confident of such an outcome, however, at the planning stage.

5. *Persist in productive endeavors.* Most good researchers work in a small number of areas at a given time, although their interests may change eventually. Conducting thematically related research is easier and more productive than conducting isolated studies across a wide range of areas and is more likely to provide answers to general, and important, questions. Switch research topics only for a good reason (e.g., available funds coupled with emerging interest), and make the initial commitment to a new area a relatively small one. If that effort is successful, subsequent commitments can grow.

Variables that Influence Design

Once one has selected a general research topic and has derived from that topic a specific experimental question, one must devise an initial strategy for addressing that question. That strategy may or may not involve true experimental methods, which always entail manipulating an independent variable and measuring the effects of this manipulation on a dependent variable. Experimentation is the best way for determining whether two variables are functionally related, but practical or ethical constraints sometimes render it impossible. Consider, for example, this experimental question: Is the age at which people first drink alcohol related to the likelihood that they will subsequently abuse the drug? One can easily envision an experiment to address this question, but conducting it would be unethical. Therefore, a researcher interested in the relation between initial drinking age and alcohol abuse would be forced to rely on correlational research. With such a strategy, pre-existing groups of subjects who began drinking at different ages, but were otherwise similar, would be compared with respect to the prevalence of alcohol abuse at a particular point in their life. Correlational and other nonexperimental strategies play a very important role in some areas of science and are occasionally useful in behavior analysis (see Chapter 6). Nevertheless, such strategies are less compelling than actual experiments and are poor substitutes for them.

There are many specific ways to conduct experiments. The term *experimental design* refers to the way in which the researcher arranges conditions so as to permit inferences about any effects that an independent variable may have on a dependent variable (cf. Johnston & Pennypacker, 1993b).

Most experiments arrange both control and experimental conditions. In the *control condition*, the independent variable is not present (i.e., is set to zero). The independent variable is present (i.e., set to some nonzero value) during the *experimental condition*, and if multiple values of the independent variable are of interest, each value constitutes a separate experimental condition. Consider a study designed to evaluate the effects of five or six doses of each of five drugs in pigeons exposed to a repeated-acquisition procedure (Picker & Poling, 1984). The control condition involved exposure to the repeated-acquisition procedure without any active drug being injected. Exposure to the repeated-acquisition procedure in the presence of each dose of every drug constituted a separate experimental condition; in all, there were 27 conditions.

There are two general kinds of experimental designs, within-subject and between-subjects. In *within-subject designs*, each subject is exposed to both control and experimental conditions, and the performance of each individual subject is compared across conditions. In *between-subjects designs*, different subjects are exposed to different conditions, and the performance of groups of subjects exposed to different conditions is compared. As a rule, within-subject designs are superior for behavior analytic research (Sidman, 1960; Johnston & Pennypacker, 1993b), although there are situations in which between-subjects designs are essential. Moreover, elements of within-subject and between-subjects designs can be combined in a single experiment, often to good avail.

Over the years, names have been assigned to a sizable number of specific experimental designs, and the general strengths and weaknesses of these designs have been discussed. For example, it is common for applied behavior analysts to differentiate and discuss separately at least four kinds of within-subject designs: withdrawal (or reversal), multiple-baseline, changing-criterion, and alternating-treatments. Such a convention, which is followed in Chapter 5. helps to facilitate communication among researchers and may help students to remember arrangements of conditions that are often useful. It is the case, however, that many good experiments combine elements of two or more specific designs. The key to a sound experimental design is to arrange conditions so that the obtained data provide a convincing demonstration of the possible effects of the independent variable on the dependent variable—not to ensure that the design can be denominated by a conventional name. There is no useful cookbook for researchers; slavish adherence to particular design conventions is a recipe for failure.

In any good experiment, conditions are arranged so that there is little possibility that the effects of the independent variable will be confused with those of extraneous variables. *Extraneous variables* are environmental events other than the independent variable that may affect the dependent variable. They pose a problem in that their effects may be *confounded* (i.e., mixed up, confused) with those of the independent variable, which may lead to erroneous conclusions concerning the effects of the independent variable.

Consider the experimental question addressed by Dickinson and Gillette (1993): How does performance under a piece-rate pay system in which 100% of the subject's pay is incentive-based differ from performance under a base pay plus incentive system? In this study, the dependent variable (i.e., performance) was the number of bank checks correctly entered into the computer in a simulated bank teller's task. The independent variable was the kind of pay system arranged. What are some potential extraneous variables in the Dickinson and Gillette study?

There are many possibilities. Among them are the age, socioeconomic status, and histories of the subject, their use of alcohol or other drugs, and the time of day that experimental sessions were scheduled. Any of these variables might influence the number of checks that a person entered.

One strategy for dealing with extraneous variables is to ensure that they are not operative in a given situation. For example, the possibility that a subject's performance on a given occasion would be influenced by drinking alcohol can be eliminated by requiring that subjects not drink prior to testing. Doing so would isolate and eliminate a potential extraneous variable (drinking alcohol). Although sound, this strategy is often difficult to utilize due to practical considerations (e.g., some subjects in this case might ignore the requirement) and the wide range of extraneous variables that might plausibly be operative in any given experiment.

A second strategy for dealing with extraneous variables is to ensure that they are relatively constant across conditions. With this strategy, the intent is to arrange comparison conditions so that they differ in only one regard: the value of the independent variable. *A cardinal rule in conducting research is to change only one variable at a time when proceeding from one phase, or condition, of a study to the next.* Consider how this strategy might deal with the effects of one potential extraneous variable faced by Dickinson and Gillette (1993), the time of day that sessions were scheduled. By arranging both pay systems during sessions scheduled at the same time each day, perhaps from 1 to 4 P.M., one can be sure that observed differences are not due to differences in the time at which the task was performed. This arrangement does not mean, however, that the obtained results would necessarily be the same if the sessions were conducted at a different time. Time of day is still a variable that could affect performance; subjects might, for instance, correctly enter fewer checks under both pay schedules if sessions were scheduled from 1 to 4 A.M. Holding extraneous variables constant across conditions does not eliminate their effects on the dependent variable, but it does help to reduce the likelihood that their effects will be mistaken for those of the independent variable.

Among the most critical requisites for research are to know the kinds of extraneous variables that are likely to pose problems in a particular experiment and to select a design that reduces to an acceptably low likelihood the possibility that the effects of those variables will be confounded with those of the independent variable. The inexperienced researcher can gain knowledge concerning the kinds of extraneous variables likely to pose a problem in particular research areas, and strategies for dealing with those extraneous variables, by reading research articles in the area of interest and consulting with established researchers. As a rule, one would do well to replicate design options that historically have proven useful in isolating the effects of particular kinds of independent variables from those of potential extraneous variables.

Of course, multiple design options are available in most research areas: The vast majority of research questions can be reasonably addressed through the use of two or more different experimental designs. The choice of design to be used in a given situation is influenced by the researcher's formal training and other experiences, available resources, ethical considerations, and the intended consumer of the information obtained.

Researchers' Training and Experience

In Chapter 1, the points were made that scientists are not fundamentally different from other people and that the same kinds of variables that influence behavior in general influence the kinds of behaviors we call "scientific." In all graduate schools, and in some undergraduate programs, students are given, and derive for themselves, rules describing gainful research tactics. They are rewarded for voicing and following these rules, and they eventually acquire a repertoire of rule-governed and contingency-shaped scientific behaviors. The nature of this repertoire depends, in large part, on the specific verbal community in which it was acquired. For example, students trained in the tradition of applied behavior analysis, perhaps at the University of Kansas, will emphasize the necessity of utilizing multiple observers and calculating interobserver agreement whenever humans are used to collect data. In contrast, calculating interobserver agreement will not be considered as important by students trained in the tradition of ethology, perhaps at Cambridge University. "Good" observational research demands a high level of interobserver agreement for those trained at Kansas, but not for those trained at Cambridge.

Events that occur after graduate training ends also influence scientists. Researchers come to use repeatedly, and to consider as appropriate, tactics and strategies that they have found to be successful, primarily in the sense of being accepted by peers. *Most researchers rely on a relatively small number of experimental procedures and conduct the majority of their studies in similar fashion, although they may be conversant with, and accepting of, other approaches.* They use particular procedures primarily because of their utility in generating data that are deemed valuable by members of a specifiable social community, for example, those who read and review papers submitted to particular journals.

Groups of scientists with common beliefs, research strategies, and conceptual frameworks are sometimes said to share a *paradigm* (Kuhn, 1970). It is tempting to attribute the common features of their actions to the paradigm, that is, to assert that scientists who behave in similar fashion do so *because they share a paradigm.* But this is circular reasoning and an error of reification. Paradigms are not things that cause people to behave in certain ways. Rather, they are shorthand descriptions of the common activities of individual members of social communities. The contingencies arranged by these communities are primarily responsible for controlling the behaviors of its members. Put simply, scientific behaviors are learned behaviors, acquired in a social context. Therefore, they may differ substantially across individual scientists.

Of course, over the long haul, different approaches to the study of behavior will not prove equally valuable in describing, predicting, and controlling what people and other animals do. The paradigm that is most effective in these regards eventually should win out and be generally accepted; the day may come when all social scientists are behavior analysts. In the interim, training and experience, and hence accepted research practices, will differ substantially across scientists.

Practical Constraints

Designing an experiment requires consideration of broad issues, such as the general arrangement of conditions that will be used. It also requires attending to many details. Table 2.1 lists several questions concerning details of procedure that must be answered before an experiment is started. The answers to these questions depend, in part, on practical considerations.

All researchers are limited with respect to available time, money, personnel, equipment, subjects, and other resources. Constraints in each of these areas influence how studies must be conducted. Consider, for instance, a scientist who is interested in using behavioral procedures to

TABLE 2.1. Questions to Answer before Beginning a Study

1. Regarding subjects:
   A. What species will be used?
   B. What special characteristics must the subjects have (e.g., particular age, gender, or psychiatric diagnosis)?
   C. How many subjects will be used?
   D. How will subjects be obtained?
   E. How long will the subjects be involved in the study?
   F. If subjects are nonhumans, how will they be housed and otherwise maintained?
   G. If subjects are humans, how will they be induced to participate?
   H. What protections for subjects are needed (e.g., project approval by appropriate Animal Care and Use or Human Subjects Insitutional Review Board committees)?

2. Regarding facilities and apparatus:
   A. Where and over what period will the study be conducted?
   B. What sorts of equipment and other materials will be used in the study, how long will they be used, and how will they be procured?

3. Regarding the experimenter:
   A. Who actually will conduct the study (i.e., serve as experimenter), and what are the necessary characteristics of the experimenter?
   B. What are the experimenter's specific responsibilities?
   C. How will the experimenter be trained and monitored?
   D. How will the experimenter be included to participate in the study?

4. Regarding the investigator:
   A. Who will be in charge of the study (i.e., serve as investigator)?
   B. How will the investigator contact the experimenter and obtain incoming data?
   C. On what basis will the investigator change the experimental protocol?

5. Regarding data collection and analysis:
   A. How will the dependent variable be defined, what dimensions of the dependent variable will be measured, and how will those dimensions be quantified?
   B. Will visual or statistical data analysis techniques be used, and what specific techniques will be employed?
   C. Who will be responsible for portraying and analyzing data?
   D. In applied studies, how will the acceptability of goals, procedures, and outcomes be assessed (i.e., how will social validity be evaluated)?

6. Regarding intervention integrity:
   A. How will the independent variable be operationally defined, and what values of the independent variable will be studied?
   B. What steps will be taken to ensure that the independent variable actually is implemented in the intended fashion?

7. Regarding problems:
   A. What difficulties are anticipated in conducting the study, and what steps are possible to obviate them?
   B. Should an anticipated difficulty arise, how will it be dealt with?
   C. Should an unanticipated difficulty arise, who should be informed and how will decisions concerning an appropriate response be made?

control food consumption in people with Prader-Willi syndrome, a rare disorder that is associated with mental retardation and severe obesity. Regardless of the locale, there are unlikely to be more than a handful of persons with Prader-Willi syndrome in the nearby area. Unless the researcher can make provision to work with subjects living far away, she or he will be forced to rely on within-subject experimental designs that can extract maximal information from a minimum number of subjects. This approach is not necessarily bad, but it does limit design options.

Successful researchers excel in utilizing available resources to optimum advantage. As they become more senior, they also excel in securing additional resources. Perhaps the best advice for the beginning researcher is to start small, using resources that you can control. Unless the resources required for a given project are at hand, or a means of securing them is readily apparent, that project is nothing more than a pipe dream. Additional resources possibly can be secured through research grants or by enlisting the aid of other scientists. Success in both endeavors increases if one can produce data demonstrating the feasibility of the proposed project; initial projects that are small but successful can lead to large opportunities. In general, those who control scientific resources give less weight to what people claim to be able to do than to what they actually have done and are currently doing.

Ethical Considerations

Obviously, any investigation must bear rigorous scrutiny with respect to the ethicality of its goals, procedures, and outcomes. Guidelines are available for the conduct of research with human subjects (e.g., American Psychological Association, 1982, 1990) and nonhuman subjects (e.g., American Psychological Association, 1985; National Institutes of Health, 1986) and should be adhered to closely. Sidebar 2-1 indicates the kinds of guidelines that are available.

Before being initiated, all research must be approved by the appropriate review boards. Securing appropriate permissions to conduct a study may entail considerable time and effort and may require modification of proposed procedures. Moreover, it is sometimes impossible to secure permission to work with particular subjects. For these reasons, it is wise to contact the appropriate review boards very early in the process of planning a study and to secure necessary permissions, including informed consents from human subjects (or their guardians), before making a binding commitment to the project.

Though ethical considerations rarely prevent scientists from conducting reasonable studies, they may limit design options. Consider a study

Sidebar 2-1
Ethical Principles of Psychologists

Principle 9: Research with Human Participants

The decision to undertake research rests upon a considered judgment by the individual psychologist about how best to contribute to psychological science and human welfare. Having made the decision to conduct research, the psychologist considers alternative directions in which research energies and resources might be invested. On the basis of this consideration, the psychologist carries out the investigation with respect and concern for the dignity and welfare of the people who participate and with cognizance of federal and state regulations and professional standards governing the conduct of research with human participants.

A. In planning a study, the investigator has the responsibility to make a careful evaluation of its ethical acceptability. To the extent that the weighing of scientific and human values suggests a compromise of any principle, the investigator incurs a correspondingly serious obligation to seek ethical advice and to observe stringent safeguards to protect the rights of human participants.

B. Considering whether a participant in a planned study will be a "subject at risk" or a "subject at minimal risk," according to recognized standards, is of primary ethical concern to the investigator.

C. The investigator always retains the responsibility for ensuring ethical practice in research. The investigator is also responsible for the ethical treatment of participants by collaborators, assistants, students, and employees, all of whom, however, incur similar obligations.

D. Except in minimal-risk research, the investigator establishes a clear and fair agreement with research participants, prior to their participation, that clarifies the obligations and responsibilities of each. The investigator has the obligation to honor all promises and commitments included in that agreement. The investigator informs the participants of all aspects of the research that might reasonably be expected to influence willingness to participate and explains all other aspects of the research about which the participants inquire. Failure to make full disclosure prior to obtaining informed consent requires additional safeguards to protect the welfare and dignity of the research participants. Research with children or with participants who have impairments that would limit understanding and/or communication requires special safeguarding procedures.

E. Methodological requirements of a study may make the use of concealment or deception necessary. Before conducting such a study, the investigator has a special responsibility to (i) determine whether the use of such techniques is justified by the study's prospective scientific, educational, or applied value; (ii) determine whether alternative procedures are available that do not use concealment or deception; and (iii) ensure that the participants are provided with sufficient explanation as soon as possible.

F. The investigator respects the individual's freedom to decline to participate in or to withdraw from the research at any time. The obligation to protect this freedom requires careful thought and consideration when the investigator is in a position of authority or influence over the participant. Such positions of authority include, but are not limited to, situations in which research participation is required as part of employment or in which the participant is a student, client, or employee of the investigator.

G. The investigator protects the participant from physical and mental discomfort, harm, and danger that may arise from research procedures. If risks of such consequences exist, the investigator informs the participant of that fact. Research procedures likely to cause serious or lasting harm to a participant are not used unless the failure to use these procedures might expose the participant to risk of greater harm, or unless the research has great potential benefit and fully informed and voluntary consent is obtained from each participant. The participant should be informed of procedures for contacting the investigator within a reasonable time period following participation should stress, potential harm, or related questions or concerns arise.

H. After the data are collected, the investigator provides the participant with information about the nature of the study and attempts to remove any misconceptions that may have arisen. Where scientific or humane values justify delaying or withholding this information, the investigator incurs a special responsibility to monitor the research and to ensure that there are no damaging consequences for the participant.

I. Where research procedures result in undesirable consequences for the individual participant, the investigator has the responsibility to detect and remove or correct these consequences, including long-term effects.

J. Information obtained about a research participant during the course of an investigation is confidential unless otherwise agreed upon in advance. When the possibility exists that others may obtain access to such information, this possibility, together for plans for protecting confidentiality, is explained to the participant as part of the procedure for obtaining informed consent.

Principle 10: Care and Use of Animals

An investigation of animal behavior strives to advance understanding of basic behavioral principles and/or to contribute to the improvement of human health and welfare. In seeking these ends, the investigator ensures the welfare of animals and treats them humanely. Laws and
(continued)

Sidebar 2-1
(Continued)

regulations notwithstanding, an animal's immediate protection depends upon the scientist's own conscience.

A. The acquisition, care, use, and disposal of all animals are in compliance with current federal, state or provincial, and local laws and regulations.

B. A psychologist trained in research methods and experienced in the care of laboratory animals closely supervises all procedures involving animals and is responsible for ensuring appropriate consideration of their comfort, health, and humane treatment.

C. Psychologists ensure that all individuals using animals under their supervision have received explicit instruction in experimental methods and in the care, maintenance, and handling of the species being used. Responsibilities and activities of individuals participating in a research project are consistent with their respective competencies.

D. Psychologists make every effort to minimize discomfort, illness, and pain of animals. A procedure subjecting animals to pain, stress, or deprivation is used only when an alternative procedure is unavailable and the goal is justified by its prospective scientific, education, or applied value. Surgical procedures are performed under appropriate anesthesia; techniques to avoid infection and minimize pain are followed during and after surgery.

E. When it is appropriate that the animal's life be terminated, it is done rapidly and painlessly.

Source: From "Ethical principles of psychologists (amended June 2, 1989)" by the American Psychological Association, *American Psychologist*, 1990, 45, 395. Copyright 1990 by the American Psychological Association. Reproduced by permission.

designed to evaluate a treatment for self-injury in children with mental retardation. Would it be appropriate to include an untreated control group in this study or to use a design in which the study ended with the withdrawal of the treatment? Probably not.

Research Consumers

Science is first and foremost a social activity. The value of research resides primarily in its ability to affect the behavior of the researcher and of other people. The people who are affected by a particular study are its consumers. The effect, if any, that a study will have depends on the question explored, the procedures employed, the results obtained, and the researcher's presentation and discussion of those results. The evaluation given to each of these dimensions for a given investigation will depend primarily on the training and other experiences of the person doing the evaluation. The sagacious researcher is aware of this dependency and considers the nature of the scientific community that will have an opportunity to respond to the project.

Consider, for example, a hypothetical student, Jan, in the M.A. program in behavior analysis at Western Michigan University. Jan's thesis advisor is Jack Michael, and the other members of Jan's thesis committee are Alyce Dickinson and Dick Malott. Each of these three faculty members is a committed behavior analyst, and each favors within-subject experimental designs, repeated and direct measures of behavioral dependent variables, and graphic (as opposed to statistical) methods of data analysis. Knowing this about the people who will evaluate the thesis, Jan would be unwise to propose a thesis testing a hypothetical cognitive model of memory processing through the use of a between-subjects design that would entail testing each of a large number of people on a single occasion. And, were the thesis completed successfully, Jan would be equally unwise to submit a manuscript based on it to the *Journal of the Experimental Analysis of Beahvior* for possible publication. Scientists who review submissions to that journal are likely to share the views of Jan's committee as to what constitutes worthwhile research and to conclude that the thesis research does not do so.

Huitema (1986b, p. 288) cautioned, "Thou shalt not commit political suicide," to remind scientists of the need to be aware of community standards when conducting and analyzing their research. It is wise to heed his advice. For example, much of our work involves assessing the effects of various drugs on the behavior of nonhumans. Most of the journals to which we submit articles for publication use many referees who favor statistical data analysis. Therefore, even though we generally favor graphic data analysis and consider it to provide a perfectly adequate means of analyzing most of our findings, we often include inferential statistics, although we do not emphasize them. Doing so increases the likelihood that our submissions will be accepted for publication, but does not substantially alter their contents or conclusions. Although one should not abandon one's own standards of quality to satisfy other people, research that is not accepted by its proposed audience is of questionable value. Unfortunately, in some cases, there is no suitable audience for research that one conducts, analyzes, and presents exactly as one wishes to, and one is forced to make some compromises. Beginning researchers have little choice save to compromise; eminent scientists may attempt to establish their own social communities (e.g., to publish new journals) or mount a head-on attack on established community standards that they view as invidious.

It can be difficult to determine the standards that a given social community uses to evaluate a research project. General standards used to evaluate submissions to a particular journal characteristically are presented therein, but the best way to discern what is required for publication in a given journal is to determine the characteristics of articles recently published in that journal. Doing so will provide reasonable clues to the kinds of experimental questions deemed of interest, the methods that are accepted, the conceptual approaches that are approved, and the kind and amount of data required. Similarly, the standards of particular persons (e.g., members of a dissertation committee or of a grant review board) usually can be determined by reading empirical articles they have published: Obviously, they deem the kind of work that they do to be "good" work. When politically appropriate, corresponding directly with potential consumers before a project is initiated can obviate many potential difficulties. This is the tack that students take when they prepare a thesis or dissertation proposal and secure the approval of their advisor and other committee members for the procedures described therein.

The Generality of Experimental Findings

Two important issues concerning the results revealed in any study are these: (1) Can similar results be reproduced under conditions essentially equivalent to those of the original study? (2) Can similar results be produced under conditions that differ in one or more aspects from those of the original study? Sidman (1960) referred to these aspects of experimental findings as *reliability* and *generality*, respectively. He emphasized that reliability, generality, and scientific importance (discussed in Chapter 3) are the key considerations in evaluating the findings of any experiment.

There is no way actually to know whether experimental findings are reliable except through conducting *direct replications*, which are studies that essentially duplicate conditions of the original investigation. A well-designed study maximizes the possibility that any revealed relation between the independent and dependent variable is real and repeatable, but error is always possible. The results of any single study must therefore be interpreted with caution. *In science, repeatability is tantamount to believability.* Relations that can be reproduced are accepted as real; those that cannot be reproduced are rejected. Science, as you will recall from Chapter 1, is not infallible, but it is self-correcting.

Once it becomes clear that particular experimental findings are reliable, the issue of generality arises: Under what other conditions can those findings be reproduced? A great deal has been written concerning the logical basis for generalizing findings from within-subject and between-subjects designs. The usual, and reasonable, argument is that between-subjects studies characteristically involve the study of a cleary defined sample of subjects who are selected so as to be representative of a larger population with definable characteristics. If this is the case, it is logical to infer that relations observed in the sample also hold in the population.

Most within-subjects experiments, in contrast, do not involve the study of clearly defined samples selected from known populations. Instead, they involve very careful study of individual subjects with specifiable characteristics, and results obtained with those subjects logically should extend to other, similar subjects.

In actuality, regardless of how a study is conducted, it is impossible to know the range of conditions under which its results can be reproduced. The only way that this range can be determined is through a series of systematic replications of the original study. In a *systematic replication*, the independent and dependent variables are similar to those examined in the original investigation, but the studies differ in at least one significant aspect. That aspect constitutes the dimension along which generality is being assessed. Behavior analysts are often interested in generality across subjects, behaviors, and settings, although these dimensions are not the only dimensions of potential importance.

As Johnston and Pennypacker (1993b) explain, it is important to differentiate speculations about generality from demonstrations of its occurrence. They write (p. 355):

> Speculations are no more than educated guesses, which are not very convincing to colleagues. Clear experimental evidence about generality is a better basis for this kind of interpretative statement. Such evidence can take different forms: (a) showing that the same results are obtained even when a certain variable is changed, suggesting that it does not limit generality; (b) showing that the results change when a certain variable is changed, suggesting that it does limit generality; and (c) systematically manipulating the variable and showing exactly how it affects the results. The third course should be most convincing. Ultimately, what we want to know is, not whether a set of results are general in some way, but what variables modulate the results and how they do so.

The variables that modulate a set of results can be determined only experimentally, and a long series of systematic replications may be required. Such a series was initiated by Herrnstein (1961), who exposed three pigeons to conditions in which concurrent variable-interval (VI) schedules of food delivery were arranged on two response keys. In most conditions, a 1.5-sec changeover delay (COD) was arranged. The COD ensured that no food could be delivered until at least 1.5 sec had elapsed from the time a bird switched from pecking one key to pecking the other. Several different VI combinations were compared. Each VI schedule specified that food became available periodically, with the average time between successive food availabilities equal to the schedule value (e.g., 60 sec under a VI 60-sec schedule), and was delivered dependent on a response. Herrnstein recorded the number of pecks emitted and the number of food deliveries obtained under each alternative. When these measures were compared across schedule combinations, the relative proportion of responses emitted under a schedule matched the relative proportion of reinforcers (food deliveries) under that alternative.

Herrnstein expressed this relation in a simple algebraic formula, which constituted the first version of the matching equation:

$$B_1/(B_1 + B_2) = R_1/(R_1 + R_2)$$

where B_1 is behavior (i.e., total responses) allocated to alternative 1, B_2 is behavior allocated to alternative 2, R_1 is the number of reinforcers received under alternative 1, and R_2 is the number of reinforcers received under alternative 2.

Subsequent studies examined whether Herrnstein's findings could be repeated in other species, with other response topographies, with other reinforcers, and with other schedules of reinforcement. Over time, the matching equation has been shown to describe accurately the relation between environmental inputs and behavioral outputs under a wide range of conditions (see de Villiers, 1977; Davison & McCarthy, 1988; Mazur, 1991), although the specific formula that describes this relation has been refined (see the section entitled "Research Topics and Experimental Questions" above) and there is some disagreement concerning the range of conditions under which matching occurs. For example, some authors have speculated that the matching law provides a useful description of the behavior of people in the workplace (Mawhinney & Gowan, 1991; Redmon & Lockwood, 1987), but others have been critical of this notion (Poling & Foster, 1993). No data directly relevant to the issue are available, however.

Although there are suggested strategies for increasing the likelihood that the results of an experiment are not limited to the precise circumstances examined (e.g., Campbell & Stanley, 1966), empirical demonstration is the only way to confirm the generality of results. Beginning researchers are well advised to recognize this fact and to be duly cautious in speculating about the generality of their findings. If generality across a particular dimension is likely to be deemed especially important by research consumers, one should attempt at least a preliminary evaluation of generality across that dimension prior to disseminating findings.

Summary and Conclusions

There is nothing especially difficult, in principle, about designing a behavior analytic experiment. The essential features of a meaningful study can, in fact, be reduced to five: (1) the experimental question must be reasonable; (2) the dependent variable must be adequately defined and measured; (3) the independent variable must be adequately defined and consistently implemented; (4) the sequencing of conditions must be adequate to confirm that the independent variable, not something else, was responsible for observed changes in the dependent variable; and (5) the results must be analyzed in a manner appropriate to determine whether the independent variable affected the dependent variable.

In reality, however, things are not so simple. Completing even the simplest study requires attention to numerous details, each of which, if neglected, can compromise the quality of the study. This chapter has emphasized some of those details and the variables that influence how they are dealt with in a particular setting and has offered practical suggestions for reducing the likelihood of significant problems. Unfortunately, no amount of forethought in designing a study can prevent problems from occurring. Research is fraught with difficulty. Accept setbacks as inevitable and press on. Successful research is a function more of dogged persistence than of perfect planning.

CHAPTER 3

Conducting and Socially Validating Experiments

Prior to starting an experiment, it is useful to develop a detailed written description of how the study will be conducted. Such a plan will provide answers to the kinds of questions listed in Table 2.1 and will describe the basis on which individual subjects will be exposed to particular conditions. The researcher must develop a detailed research plan in order to ascertain necessary resources and to obtain permission from research review committees to conduct a study. Such a plan also helps to ensure that the conditions arranged in a study actually are comparable to those intended by the person who designed the study.

Applied behavior analysts characteristically go to considerable lengths to ensure that their dependent variables are appropriately defined and measured and that experimental designs are sufficient for determining whether or not observed changes in target behaviors can be attributed to a given intervention. Historically, they have been less concerned with adequately defining their independent variables (interventions) and demonstrating that these interventions are consistently applied in the manner intended by their designer (Peterson, Homer, & Wonderlich, 1982; Poling, Smith, & Braatz, 1993). If they are not, the researcher is not actually evaluating the independent variable of interest and cannot accurately answer the experimental question. Moreover, if an intervention is described to research consumers in its intended (not actual) form, it is impossible for the reader of an article to know what was responsible for observed outcomes, to replicate the study, or to build on or refine what may appear to be a budding technology. In view of these potential problems, cavalier disregard of the integrity of interventions is unwise (Barber, 1976; Peterson et al., 1982; Poling et al., 1993).

Determining Intervention Integrity

The general strategy for examining the integrity of independent variables is to have someone who knows what should be done evaluate what actually is done. This evaluation can involve informal rating or formal rating that provides a statistical estimate of the reliability of treatment implementation (Peterson et al., 1982). Direct observation of the implementation of a procedure is desirable, but informal descriptive checks are better than nothing. In some cases, evidence of the integrity of the independent variable can be determined and reported with relative ease. For example, Henry and Redmon (1990) described in a brief paragraph how they ascertained that their intervention was implemented as intended (pp. 34–35):

> The quality control manager spoke very little to the operators concerning their daily performance at the time the cards were distributed. To insure this, and to make certain that the subjects consistently came in contact with the independent variable, a researcher often accompanied the manager. In forty percent of the measurement sessions, a researcher observed from a distance as the manager gave the index cards to the operator. No observations were recorded in which the manager failed to give a card to an operator who was supposed to receive one.

Their intervention involved providing feedback to operators via manager-distributed cards, and the aforecited passage helps to assure readers that the cards were distributed as intended by Henry and Redmon and described in their article.

In some cases, those who implement an intervention change and refine it over time; therefore, probes evaluating the integrity of an intervention need to be scheduled throughout the course of an investigation. Such probes may shed light on why a particular nominal intervention actually was successful (or unsuccessful) and also suggest variables for future research.

The integrity of the independent variable cannot be assumed in any study, even in laboratory experiments in which computers or other machines control conditions. Programming errors occur and machines malfunction, making it obligatory to determine empirically that the conditions to which subjects are exposed are precisely the same as those actually arranged. This determination can be made by having the experimenter assume the role of subject and actually respond under the conditions of interest. Good researchers who work with nonhuman subjects "play pigeon" on a regular basis.

Even when machines are working properly, humans can use them in the wrong way. Providing specific written guidelines for what the experimenter is to do each day can reduce experimenter error in basic and applied studies alike. Guidelines such as those in Sidebar 3-1, which outline the operation of a computer used to control studies in our laboratory, are valuable job aids, although they are not substitutes for adequate training and monitoring of personnel.

Flexible Designs

In many studies, it is not possible during the design phase to specify exactly the conditions to which particular subjects will be exposed or the duration of their exposure to those conditions. The manner in which behavior analytic research is conducted characteristically depends, in large part, on incoming data (Sidman, 1960; Johnston & Pennypacker, 1993b). Consider, for example, a study conducted by Ragotzy, Blakely, and Poling (1988). The purpose of the study was to examine in largely nonverbal people with mental retardation whether the time at which a choice is given relative to the availability of two alternative reinforcers influences preference for the larger, but more delayed, of those reinforcers. Self-control has been defined by several researchers as choosing of the larger, but more delayed, of two reinforcers (e.g., Ainslie, 1974; Green, Fisher, Perlow, & Sherman, 1981; Rachlin & Green, 1972), and there has been substantial interest in isolating the variables that determine whether such behavior occurs.

As a result of reading the relevant literature and having experience with similar procedures and subjects, we were able to specify in considerable detail the precise experimental procedures to be used and the temporal sequence in which participants would be exposed to those procedures. In brief, each experimental session involved 10 forced-exposure and 10 choice trials. In a forced-exposure trial, either a yellow circle or a purple triangle was presented, and the subject was asked to touch it. Doing so resulted in delivery of either one or three Cocoa Puffs (a chocolate-flavored breakfast cereal), depending on whether the stimulus presented and touched was the circle or the triangle. In a choice trial, the circle and triangle were presented together, and the subject was asked to touch one of the forms. Doing so resulted in delivery of the reinforcer correlated with that stimulus during forced-exposure trials.

Pilot Studies

Although we were reasonably confident that this general procedure would work, we tested its adequacy in a pilot study involving two subjects similar to those to be used in the actual experiment. A *pilot study* is an

Sidebar 3-1
A Sample Job Aid

It is imperative that experimenters have clear instructions to follow to ensure that an experiment is conducted as intended by the investigator. Presented below is a job aid that we use in our laboratory to train students with no experience in operating laboratory equipment used in nonhuman research. One of the functions of the job aid is to provide answers to common questions. Such a job aid reduces the problem of students having questions and not being able to contact their supervisor for an answer. Moreover, the supervisor is relieved of the burden of repeatedly having to answer simple questions.

Instructions for Conducting Experiments

1. To activate the system:

Turn power on at the outlet strip (A) on the side of the interface cabinet. This will supply power to (turn on) the interface (B), noise generator (D), and the transformer (E).

Turn on the computer. The on/off switch (F) is in the upper right corner of the front of the computer. The monitor automatically turns on when you turn on the computer.

Turn on the printer at switch G.

As the computer boots up, you will notice various system messages on the monitor (e.g., virus protection). The system check takes about one minute, after which the C:> prompt appears. At this point, you must change to the MED-PC directory by typing CD MED-PC (space between CD and MED). The C:\MED-PC> prompt will appear. You may now load your program.

2. To load a program:

Type "Dan" and hit return. This is called your "batch job" and loads the MED-PC system with any programs that were listed when the batch job was created.

You may load your particular programs in one of two ways. The easiest way is to run what's called a "macro." A macro is a command created by the user to execute a certain sequence of keystrokes. A macro is executed by pressing F4, typing the name of the macro at the "Macro to play" prompt, and hitting RETURN. The macros you can play and their respective functions are as follows:

Macro	Function
PR5M1	Loads the PR5 program into boxes 1–4 for the first group
PR5M2	Loads the PR5 program into boxes 1–4 for the second group

3. To run a program:

Once the program is loaded into the boxes, the box check must be done. Hit the center key in each box. This will turn on the relevant outputs used in the program (houselight, feeder, feeder light, and white light on the center key). Check to see that all of these outputs are activated. Also, make sure that there is enough food in the feeders. You should be able to touch the grain inside the feeder with your middle finger. This is the half-full mark (more or less). DO NOT FILL FEEDERS MORE THAN HALF FULL. Otherwise, the feeder will not operate correctly (i.e., will be too slow).

Hit the center key again to turn all the outputs off. You are now ready to start the program.

To start the program, hit "S" followed by "1,2,3,4," and then hit RE-TURN.

The session is over when the box numbers on the lower-left half of the screen are flashing

At this point, you can load the program for the next group of birds by following the macro instructions above or, if it's the last group for the day, hit "Q" to exit MED-PC.

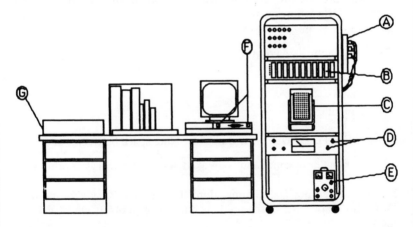

A. Power (outlet) strip.
B. MED-PC interface.
C. MAC Panel: Make sure that this panel is engaged before you do the box check, otherwise nothing will work. The panel is engaged when the metal lever is in the down position. Also make sure that the appropriate panel is in place. You need to use the panel labeled "Rat Boxes."
D. Noise generator: Make sure that both switches are in the up position, the noise is set on "white" (not "speech"), and the meter is reading −5db (this is adjusted by turning the "gain" dial).

(*continued*)

Sidebar 3-1
(Continued)

E. Power transformer: Make sure both switches are in the up position and the large dial is turned all the way to the right.
F. Computer on/off switch (right) and reset switch (left)
G. Printer on/off switch

Note: You shouldn't have to mess with B, D, and E. However, if anything goes wrong, check to make sure those components are adjusted properly (as described above).

4. To retrieve data from a session:

Once you have quit the last session of the day (i.e., hit "Q" to return to the MED-PC prompt), you must change directories to retrieve your data. To do this, type CD DATA (space between CD and DATA). The result is a new prompt, C:\MED-PC\DATA>. You are now in the data subdirectory of the MED-PC directory. You may now run your data analysis program.

To run the data analysis program, type MLDATA at the C:\MED-PC\DATA prompt (the program will not work if you are in a different directory). The following will occur. Anything below that is in bold print indicates a response typed by you.

How many animals were run?
8   ←   After each entry, hit RETURN.
Name of input file?
!930617
Name of output file?
Dan
C:\MED-PC\DATA>**Print Dan**   →   Note space between t and D.

When the print command is invoked, the printer will start printing the data.
NOTE: If anything goes wrong, don't try to fix it. Just give me a call (373-6610). If you smell or see anything burning or smoking, *immediately* shut the system down.

---

experiment conducted prior to a major research effort either to test the adequacy of particular aspects of procedure or to explore informally whether two variables are functionally related (cf. Sidman, 1960; Johnston & Pennypacker, 1993b). Although a pilot study is not intended to explore fully the relation between an independent and a dependent variable, or to provide an adequate answer to an experimental question, it must be conducted carefully, with the same attention to detail as in an actual experiment. As Sidman (1960, p. 218) emphasized, "If a pilot study is not run under exactly the same conditions as would be required in a full-scale experiment, its predictive value is completely negated." Ergo, it is without worth.

Even when pilot work suggests that the technical aspects of a proposed procedure will prove adequate, it may be impossible to specify before the fact the exact values of the independent variable that will need to be examined in each subject or the length of time that subjects will be exposed to each value. This impossibility obtained in the Ragotzy et al. (1988) study. To ensure that all subjects would be treated similarly and that our procedures could be readily described to other researchers, we established criteria for determining the range of values of the independent variable that would be examined and the length of exposure to each variable. In the first phase of the study, each of three subjects was allowed to choose either one or three Cocoa Puffs. Initially, each reinforcer was presented immediately. When responding stabilized, delivery of the larger reinforcer was delayed by 5 sec. Throughout the study, the criterion for stability was three consecutive sessions in which the percentage of choices of the larger reinforcer did not vary or five consecutive sessions in which this measure varied by less than 10%.

In subsequent conditions, the delay was increased in multiples of 5 sec until the percentage of choice responses directed to the stimulus correlated with the larger reinforcer was 20% or below on three consecutive sessions. This cutoff percentage was reached with the larger reinforcer delayed 10 sec for Subject 1, 20 sec for Subject 2, and 30 sec for Subject 3. In the balance of phase 1, the delays already examined were arranged in descending order.

In the initial condition of phase 2, the smaller reinforcer was delayed by 0 sec and the larger reinforcer was delayed by the value that shifted preference from the larger to the smaller reinforcer (e.g., 10 sec for Subject 1). In the next condition, 5 sec was added to the delay for each reinforcer. Thus, for Subject 1, the larger reinforcer was delayed by 15 sec and the smaller by 5 sec. Each subsequent condition added 5 sec to the delay of each reinforcer. For Subjects 1 and 2, delays were increased until over 80% of choice responses were directed to the larger reinforcer on three consecutive sessions. This point was reached for Subject 1 with the smaller reinforcer delayed 15 sec and for Subject 2 with the smaller reinforcer delayed 25 sec. For Subject 3, delays were increased to 30 sec for the smaller reinforcer and 60 sec for the larger. Because considerable emotional re-

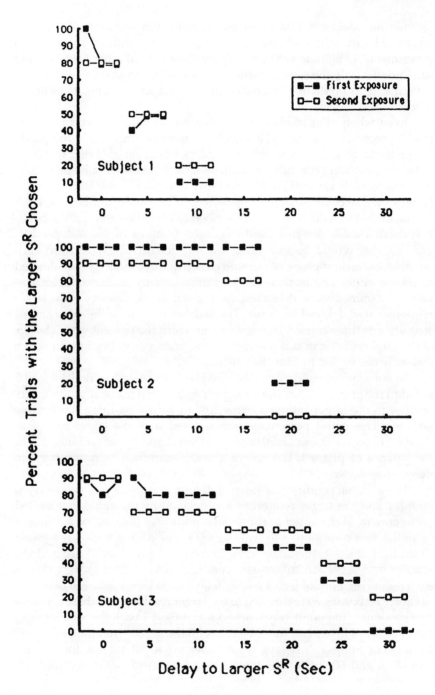

sponding was observed at this value and the school year was ending, no further increases were arranged, although the criterion for preference had not been met. After the ascending series of delays was evaluated, the same delays were examined in a descending sequence with Subjects 1 and 2, with the exception that the largest delay was repeated at some point. In order to complete the study during the school year, for Subject 3, delays were decreased by multiples of 10 sec during the descending series.

Results of the study are shown in Figures 3.1 and 3.2. In brief, under conditions in which the smaller reinforcer was not delayed, increasing the delay to delivery of the larger reinforcer decreased the percentage of trials in which that reinforcer was chosen. All subjects directed the majority of choice responses to the smaller reinforcer when the larger reinforcer was sufficiently delayed, although the value at which this redirection occurred differed across subjects. Under conditions in which the larger reinforcer initially was sufficiently delayed to result in preference for the smaller one, progressively increasing the delay to both reinforcers in 5-sec increments increased the percentage of trials in which the larger reinforcer was chosen. At sufficiently long delays, two of the subjects consistently chose the larger, but more delayed, reinforcer, and the third subject chose that reinforcer in half the trials. These results are consistent with the findings of prior studies in which adult humans who were not mentally retarded responded to terminate noise (Millar & Navarick, 1984; Navarick, 1982; Solnick, Kannenberg, Eckerman, & Waller, 1980) or in which pigeons responded to obtain food (Ainslie, 1974; Rachlin & Green, 1972; Green et al., 1981), and they demonstrate a form of self-control in largely nonverbal people with mental retardation.

The study by Ragotzy et al. (1988) was described in considerable detail to emphasize, first, that it is not necessary in behavior analytic research to treat all subjects in precisely the same manner and, second, that having clear decision rules for the treatment of all subjects facilitates both conducting and describing studies. Good behavior analytic research is flexible, but it is not capricious. *The simplest arrangement of conditions that provides a compelling answer to the experimental question is likely to be best. Treat different subjects differently only when there is a defensible reason for so doing.*

←_____

FIGURE 3.1. Percentage of trials in which each of three mentally retarded adolescents chose three pieces of breakfast cereal over one piece during the last three sessions of each condition of phase 1. During this phase, the smaller reinforcer was not delayed and the larger reinforcer was delayed by the value indicated on the abscissa. Data for each session represent 10 choice trials. From Ragotzy et al. (1988, p. 196). Copyright 1988 by the Society for the Experimental Analysis of Behavior. Reproduced by permission.

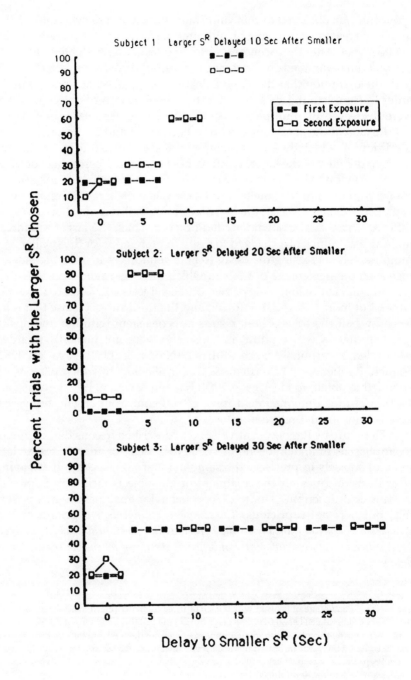

Subject 1   Larger S^R Delayed 10 Sec After Smaller

First Exposure
Second Exposure

Subject 2: Larger S^R Delayed 20 Sec After Smaller

Subject 3: Larger S^R Delayed 30 Sec After Smaller

Percent Trials with the Larger S^R Chosen

Delay to Smaller S^R (Sec)

Extreme Values and Problems of Generalization

It is a common strategy, especially in basic research, to determine the range of values of the independent variable to be examined in a research project on the basis of the behavioral effects observed at particular values. Behavioral pharmacologists, for instance, often evaluate drug doses ranging from those too small to produce a behavioral effect to those large enough to nonselectively suppress responding (Poling, 1986). When one does not know the appropriate dosages, a reasonable strategy is to initially evaluate a very small dose, then systematically increase the dosage (perhaps in quarter-log or half-log units) until clear effects are produced. If desired as a partial control for order effects, these doses can then be reexamined when administered in random order.

The concept of dose-dependent effects is fundamental to all of pharmacology; thus, it is essential to evaluate several doses in any investigation. The dosing procedure just outlined is practical and provides a means of evaluating drug doses between the behaviorally inactive and clearly toxic levels. But it poses a significant problem concerning the generality of results: When values of the independent variable outside the range commonly encountered in the situation of eventual concern are evaluated in an experiment, generalization to the situation of concern is tenuous. Recall our studies of the behavioral effects of anticonvulsant drugs in nonhumans discussed in Chapter 2. We used rats and pigeons as subjects and gave drugs by injection (humans with epilepsy take them by mouth), and the doses required to produce behavioral effects were substantially larger than those that are used clinically. Thus, a reasonable criticism of our work was that we were studying excessively high doses, making our findings irrelevant to the effects of anticonvulsants in humans with epilepsy.

To evaluate this possibility, we evaluated anticonvulsant and behavioral effects in the same animals (Renfrey et al., 1989) and found that the doses that controlled seizures also disrupted performance. Thus, the doses we were evaluating were not unreasonably large. Unfortunately, in many research areas, there are no obvious benchmarks for determining appro-

---

FIGURE 3.2. Percentage of trials in which each of three mentally retarded adolescents chose three pieces of breakfast cereal over one piece during the last three sessions of each condition of phase 2. During this phase, the smaller reinforcer was delayed by the amount indicated on the abscissa and the larger reinforcer was delayed by that value plus the value indicated at the top of each graph. Data for each session represent 10 choice trials. From Ragotzy et al. (1988, p. 49198). Copyright 1988 by the Society for the Experimental Analysis of Behavior. Reproduced by permission.

priate values of the independent variable. In such cases, perhaps the best advice is to evaluate a wide range of values of the independent variable, provide consumers with precise information about how those values were selected, and exercise extreme caution in speculating about how results generalize to other situations. Examining extreme values of the independent variable may maximize the likelihood of showing a functional relation between that independent variable and the dependent variable of interest, but the relation may or may not hold with other values. As emphasized in Chapter 2, the only way to tell is through systematic replication.

Pursuing Serendipitous Findings

Occasionally, interesting behavioral phenomena not directly anticipated in the experimental question will become apparent to the researcher who closely monitors incoming data. In such cases, should the initial experimental question and correlated design strategy be abandoned to allow the researcher to purse such a serendipitous finding? There is no straightforward answer. Many important discoveries have arisen from the actions of scientists who noticed unplanned relations between variables, then took steps to clarify the nature of those relations. Certainly there is precedent and logical justification for changing the focus of an experiment in midcourse. If one does so, however, it is important to ensure that the conditions arranged are adequate for providing useful information about the variables of interest. In many cases, it is better to continue an experiment as planned and to devise a new study to explore the relationship revealed by accident.

Occasionally, it becomes clear in the course of conducting a study that the study is seriously flawed and will not yield meaningful information. This situation can arise when, for example, a previously unforeseen extraneous variable is confounded with the independent variable, the independent variable is not implemented as designed, great variability is evident in the data, or subject attrition is excessive. At the point that the investigator decides that an experiment has a serious flaw, it becomes, at best, a pilot study. As Sidman (1960, p. 219) explains, "Its data are useful only insofar as they have revealed the technical inadequacy, although they may also contain hints as to the means for rectifying the fault." Although careful planning can substantially reduce the number of unintentional pilot studies, such studies are an inevitable part of all research programs. Productive scientists learn from them; unproductive scientists whine and quit.

Social Validation

Sidman (1960) emphasized that scientific findings merit examination in three regards—specifically, with respect to their reliability, generality, and scientific importance. As discussed in Chapter 2, the reliability of findings is determined through direct replications and their generality through systematic replications. Over the short term, the scientific importance of findings is determined by the degree of enthusiasm with which a collective of scientists responds to the results of a particular experiment. If, for example, an article submitted for possible publication in the *Journal of the Experimental Analysis of Behavior* is lauded by the reviewers, promptly accepted for publication, and published as the lead article, the reported findings are considered to be important by at least some scientists concerned with the experimental analysis of behavior.

The scientific importance of experimental findings is neither absolute nor everlasting. The findings of a very large number of studies have played some role in advancing our understanding of the variables that control behavior in humans and other animals. But only a very much smaller number of studies have yielded results that substantially influenced either the development of the science of behavior analysis or the culture at large. History affords the only perspective that allows one to determine whether particular findings are especially important. Interestingly, scientists often produce important findings early in their careers, as Sidman (1960, p. 41) emphasized:

> The cumulative development of a science provides the final answer to the importance of any particular data; sometimes it is the younger scientists, who enter the field unencumbered by the prejudices of past controversies, who pick out the threads of continuity from the tangle of theory, data, and pseudo-problems that form a part of every stage of scientific progress.

The reliability, generality, and scientific importance of findings are important in applied as well as basic research areas. In applied research, the social importance of findings is also of concern.

Behavior analysts use the term *social validation* to refer to methods of determining the social significance of the goals, appropriateness of the procedures, and social importance of the results of an intervention (Kazdin, 1977; Wolf, 1978). Geller (1991) distinguished the significance of goals, appropriateness of procedures, and importance of outcomes as follows: A socially significant goal represents a deficit in functioning as society views it (e.g., a lack of appropriate social skills). An appropriate procedure is one that produces minimal adverse effects (e.g., a drug reduces aggressive

behavior without producing gross sedation). A socially important outcome is one that enhances subjects' functioning in their environment (e.g.,
a disabled person becomes competent in the use of public transportation).

Social validation may be accomplished by subjective evaluation or
social comparison (Kazdin, 1977). These methods are often used separately, but they can be combined. With the *subjective evaluation* method,
the goals, procedures, and outcomes of an intervention are evaluated by
the subject, people who interact with the subject on a regular basis (e.g.,
parents, teachers), or experts in an area relevant to the subject's problem.
With the *social comparison* method, the behavior of a subject before and
after treatment is compared to the behavior of "normal" peers. Presuma-
bly, the subject's behavior prior to treatment is significantly different from
the behavior of peers who are considered by society to be functioning
normally in their environment and thus do not require treatment. An
intervention that brings a subject's behavior within the range of his or her
peers' behavior would constitute a socially important effect, because the
subject's behavior after treatment would be indistinguishable from that of
normal peers.

Subjective Evaluation

The main task of subjective evaluation is to ask consumers of an
intervention how satisfied they are with the goals, procedures, and outcomes of the intervention. Consumers are the people who are affected
either directly or indirectly by an intervention; they might include the
subjects themselves, parents, guardians, teachers, direct-care staff, and
advocates. In essence, consumers are asked whether the appropriate problems were targeted, the procedures were acceptable, and the intervention
"made a difference" in the subjects' level of functioning. Subjective evaluations are typically obtained through interviews, questionnaires, rating
scales, or surveys. Some examples of scales that are used to assess consumer satisfaction are the Parents Consumer Satisfaction Questionnaire
(PCSQ) (Forehand & McMahon, 1981), the Behavior Intervention Rating
Scale (Elliot & Von Brock Treuting, 1990), the Children's Intervention Rating Profile (Witt & Elliott, 1985), and the Treatment Evaluation Inventory
(TEI) (Kazdin, 1980).

Consumer satisfaction scales are often modified to suit the researcher's specific needs. For example, Dachman, Alessi, Vrazo, Fuqua,
and Kerr (1986) evaluated the effectiveness of a multicomponent training
program to teach infant care skills to first-time fathers. To assess their
satisfaction with the training program, each subject and the subject's wife
completed a modified form of the PCSQ. Table 3.1 shows a small sample of

TABLE 3.1. A Sample of Items from a Modified Form of the Parent's Consumer Satisfaction Questionnaire[a]

1. At this point, my expectation for a satisfactory outcome of the training program is:

    1 = very pessimistic   7 = very optimistic

    Response: 6.0 (fathers)   6.0 (wives)

2. I feel the approach to teaching infant care by using this type of training program is:

    1 = very inappropriate   7 = very appropriate

    Response: 6.0 (fathers)   5.75 (wives)

3. Would you recommend the program to a friend or relative?

    1 = strongly not recommend   7 = strongly recommend

    Response: 6.0 (fathers)   5.75 (wives)

4. How confident are you in your ability (your husband's ability) to use these infant care skills with your newborn?

    1 = very unconfident   7 = very confident

    Responses: 6.6 (fathers)   6.75 (wives)

5. My overall feeling about the training program is:

    1 = very negative   7 = very positive

    Response: 6.0 (fathers)   6.25 (wives)

6. At this point, I think my husband's ability to handle caretaking concerns is:

    1 = considerably worse   7 = greatly improved

    Response: 6.25

[a]From Dachman et al. (1986).

the items used. Participants responded to each question by specifying a number between 1 and 7. Descriptors were provided for the 1 and 7 endpoints of the scale. Numbers indicate the mean responses to the questionnaire for fathers and wives. Note that the questions from the PCSQ were specifically tailored to the target behaviors (infant care skills) and intervention (training program) employed in that study.

Examples of Subjective Evaluation

Miller and Kelly (1994) used subjective evaluation methods to evaluate the social validity of using goal setting and contingency contracting to improve children's homework performance. Parents were administered an adapted version of the PCSQ to provide ratings of their satisfaction with the overall program, teaching methods, treatment procedures, and therapist behavior. Total scores on the PCSQ could range from 20 to 140, with higher scores representing greater satisfaction. Parents and children were also given unstructured follow-up interviews to assess their perceptions of the treatment. They were asked to identify strengths and weaknesses in the procedures and positive and negative effects. The scores obtained on the PCSQ ranged from 107 to 134. In general, parents reported that goal setting and contingency contracting were highly useful and fair procedures that were easy to implement. Parents also reported satisfaction with their child's progress with homework. Thus, the procedures and outcomes appeared to be socially validated.

A study by Kennedy (1994) provides a good example of employing people with expertise, as opposed to consumers, to evaluate subjectively the social validity of the procedures and outcomes of an intervention. The effects of instructor's task demands and social comments on problem behaviors such as stereotypy, self-injury, and aggression were evaluated in three students with severe disabilities.

Uninformed personnel from classrooms for students with severe disabilities and from a supported-employment organization (i.e., people who had experience working with handicapped students, and thus some degree of expertise) assessed the importance of intervention outcomes by responding to four questions from the Motivation Rating Scales (Dunlap & Koegel, 1980; Koegel & Egel, 1979). The questions referred to the students' general behavior, interest, happiness, and productivity. Each question was scored on a 6-point scale with descriptions appropriate to the question anchoring the end- and midpoints. Raters viewed four 30-sec videotapes of each student taken during baseline conditions and four 30-sec videotapes taken during the final days of the intervention. Baseline and intervention tapes were presented in a mixed order so that raters would be unaware of the condition being viewed. Raters detected positive changes in all subjects across all four categories.

The same raters also assessed the acceptability of the procedures by responding to eight questions from the TEI (Kazdin, 1980). Each question was scored on a 7-point scale, with descriptive anchors at the end- and midpoints. Raters received a verbal description of the intervention and then viewed four randomly selected 30-sec videotapes of each student taken at various points during the course of treatment. Raters reported that the intervention was an acceptable means of treating problem behavior and one that they would be willing to use.

Concerns with Subjective Evaluation

Although subjective evaluation methods have proven useful to behavior analysts in socially validating the goals, procedures, and outcomes of an intervention, there are three important issues to consider when evaluating the data obtained through such methods. One is that subjective evaluations may be inaccurate to the extent that the consumer's evaluation is inconsistent with actual changes in the subject's behavior, or unreliable to the extent that the consumer's evaluation is inconsistent across assessments.

The study by Miller and Kelly (1994) discussed above provides an example of how the accuracy and reliability of consumers' evaluations can be assessed. Recall that parents were administered an adapted version of the PCSQ to obtain ratings of their satisfaction with an intervention designed to improve their children's homework. In addition to the PCSQ, parents completed a Homework Problem Checklist (HPC) to measure changes in homework. Despite their positive evaluation of the intervention on the PCSQ and the improvements in measures of homework accuracy and time on task, parents' perceptions of troublesome homework-related behaviors did not change, as indicated by their ratings on the HPC. Thus, the accuracy and reliability of parents' evaluations on the HPC (and hence the social validity of the intervention) are questionable in that they are inconsistent both with observed increases in homework accuracy and time on task and with parents' own positive evaluations of the intervention on the PCSQ.

A second issue relevant to the use of subjective evaluation is that high levels of consumer satisfaction do not necessarily indicate that there has been a clinically important change in the subject's behavior (Kazdin, 1982). It is important to realize that a consumer's evaluation of an intervention via a questionnaire or an interview constitutes a type of self-report data. Bias is always a concern in evaluating self-report data (see Chapter 4); therefore, one needs to be cautious in evaluating a consumer's subjective evaluation of an intervention. Consumer opinion may be biased for several reasons, including an expectation that an intervention will work or a failure to attend to the aspect of the subject's behavior that the intervention actually is intended to change. A truly effective intervention produces desirable changes in the aspect of the subject's behavior targeted for change and satisfies consumers by so doing.

A third issue in pursuing social validation through subjective evaluation concerns determining who actually constitutes the "consumers" of an intervention. Consider the use of psychotropic drugs for treating behavioral problems in people with mental retardation. As Poling and LeSage (1995, p. x) indicate:

> An especially important issue here concerns the extent to which persons with mental retardation can meaningfully evaluate the goals, procedures, and results of pharmacological interventions. With most adults who are not mentally retarded, the ultimate form of social validation is that they choose to initiate and continue exposure to a particular medication. Many mentally retarded people do not have the capacity to make such choices, and their consent to receive medication is by proxy. In such cases, it is crucially important that other people with the patient's best interests in mind are well satisfied with the goals, procedures, and outcomes of drug treatment. Such individuals, foremost among them parents, other legal guardians, and advocates, appear to be the best sources of social validity data.
>
> But even equally concerned people may disagree in their evaluation of a pharmacological intervention, which constitutes a second potential problem. A third [discussed earlier in the paper] is that social validity data may not parallel direct measures of target behavior. Parents may, for instance, report that naloxone is not acceptable as a treatment for their child, even though the drug reduced self-injury substantially and produced no adverse reactions that could be detected via conventional side effects rating scales.

A reasonable tack to take when social validity data obtained through subjective evaluations by different people are dissimilar, or when such data are not supported by actual measures of the subject's behavior, is to determine the reason for the discrepancy. If a researcher is unable or unwilling to take this step, probably the best strategy is simply to present all the data collected and indicate that the apparent value of the intervention is dependent on how, or by whom, it is evaluated.

Social Comparison

Use of the social comparison method begins with identification of the subject's peer group. The members of the peer group should be similar to the subject in important characteristics such as age, gender, and socioeconomic status, but differ from the subject with respect to the target behavior. The level at which the peer group exhibits the target behavior should be that which society considers acceptable or "normal." For example, distraught parents may seek treatment for their 5-year-old boy, Tommy, who teases his siblings at an average rate of 50 times per day. To what level would an intervention need to decrease this behavior in order for the effects to be considered socially important? To find out, one could identify Tommy's peer group—boys similar to Tommy in age, general functioning level, and socioeconomic status—and measure how often they tease their siblings. Perhaps they do so twice a day, which constitutes a criterion level relative to which Tommy's behavior can be evaluated. If his teasing is reduced to two or fewer occurrences per day, such an outcome is socially important because his behavior in the area of concern is no longer different from that of his peers.

Examples of Social Comparison

Behavior analysts have used social comparison methods both to identify behaviors selected for change (target behaviors) and to evaluate the social importance of treatment effects. For example, Lagomarcino, Reid, Ivancic, and Faw (1984) evaluated a training program to teach institutionalized severely and profoundly mentally retarded persons how to dance. The specific dance skills taught were based upon observations of the dance behaviors of noninstitutionalized mildly and moderately mentally retarded peers and nonretarded peers of approximately the same age as the subjects. After the training, the dancing skills of the subjects were within the normative range (i.e., like those of their peers); thus, the intervention produced socially important effects.

Another example of the use of the social comparison method is a study by Gajar, Schloss, Schloss, and Thompson (1984). They evaluated the effectiveness of feedback and self-monitoring in increasing the positive conversational skills of youths with head trauma. To assess the importance of treatment effects, a social comparison method was used. With this method, the behaviors targeted for change in the subjects were measured in untreated peers (two groups of six 20- to 22-year-old college students) in the same situations to which the treated subjects were exposed. In one situation (treatment), the facilitator read a "Dear Abby" article to provide a topic of conversation. In the other situation (generalization), the facilitator simply asked the participants, "What shall we talk about today?"

Figure 3.3 shows the percentage of appropriate conversational behaviors for one of the subjects. The figure also shows dashed horizontal lines that indicate the mean and standard deviation of peer behaviors. These lines make it easy to assess whether the subject's behavior was brought into the normative range of functioning. During baseline phases (A1 and A2), the subject's conversational behavior was below the norm (i.e., below the lowest dashed line, which represents the low end of the normative range of conversational behavior). During every session of the intervention phases, appropriate conversation increased to within the normative range. Thus, the treatment effect was socially important.

Concerns with Social Comparison

Researchers who contemplate the social comparison strategy face two potentially vexing issues. One is that it may be difficult to choose an appropriate comparison group for a particular subject. For example, to

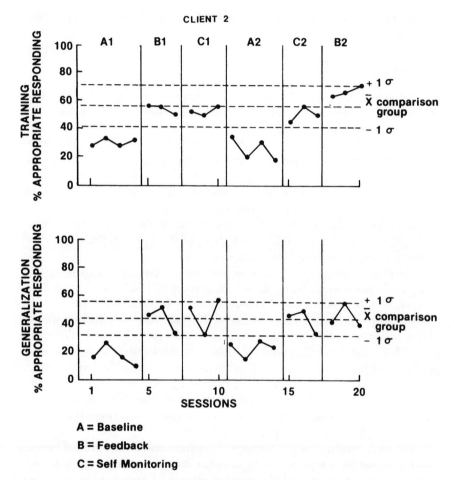

FIGURE 3.3. Percentage of appropriate conversational behaviors produced by one client during baseline, treatment, and generalization sessions. The client had suffered head trauma and was exposed to a treatment involving feedback and self-monitoring. From Gajar et al. (1984). Copyright 1984 by the Society for the Experimental Analysis of Behavior. Reproduced by permission.

whose behavior should the locomotor behavior of a physically challenged person be compared? Normative levels of locomotion of people who are not physically challenged may represent an unrealistic target for a physically challenged person. Yet the physically challenged person may be capable of locomotor behavior more nearly normative than that of similarly challenged people who have not received the best possible treatment. Here, the appropriate comparison group appears to be similarly challenged peers who have received excellent treatment, but such a group may be difficult to define or access.

The possibility that normative levels of behavior are undesirable is a second issue to consider when using social comparison methods. If normative levels are inappropriate, they are not sound goals for treatment, and reaching them does not constitute good evidence of the social importance of obtained effects. Assume, for example, that a seventh-grade student of average intelligence is taught to read as well as her or his classmates—all of whom read at the second- or third-grade level. That student still has a reading problem by any reasonable standard. To counter such problems, Van Houten (1979) has suggested that treatment goals be based on the performance of persons considered as competent, not simply as representative, and has described empirical procedures for determining the optimal level of a given behavior.

Summary and Conclusions

Behavior analysts characteristically use flexible experimental designs and base many decisions about how an experiment should be conducted on incoming data. Although logic and pilot studies provide valuable clues as to how a study should proceed, *let the data be the guide*. This approach to research has much to recommend it, including the capacity to explore unanticipated, but interesting, behavioral relations (Barlow & Hersen, 1984; Johnston & Pennypacker, 1993b; Sidman, 1960).

To prevent flexibility from becoming chaos, be sure to base decisions about the appropriate treatment of individual subjects on defensible, and readily describable, criteria. In developing such criteria, keep the intended consumer of the research in mind and remember the general rule that simple experimental arrangements are best.

Few studies go smoothly; most are flawed to a degree. That a study is not above criticism does not mean, however, that its findings are of no value. According to Sidman (1960), the worth of a study to the science of behavior analysis depends on the reliability, generality, and scientific importance of its findings. For applied studies, social importance is also significant.

Social validation methods are useful for evaluating the social importance and acceptability of the goals, procedures, and outcomes of an intervention, although such methods may only give rise to problems, such as subjective biases and unacceptable normative levels of behavior. Potential problems notwithstanding, social validation has the virtue of including people other than the researcher in the evaluation process (Fuqua & Schwade, 1986). As Wolf (1978, p. 207) suggests, the employment of social validation methods might "bring the consumer, that is society, into our science [applied behavior analysis], soften our image, and make more sure our pursuit of social relevance."

Collecting and including social validity data adds little to the cost or difficulty of conducting a study or the space required to report it. Doing so, however, allows readers to respond to the other data reported in a broader, and more socially significant, context. For this reason, *we recommend that researchers working in applied areas routinely collect and report social validity data, as well as objective measures of target behaviors. We also recommend that researchers in both basic and applied areas always take steps to ensure the integrity of their independent variables.* The consequences of failing to do so truly can be calamitous.

CHAPTER 4

Data Collection

The experimental question generally dictates the specific behaviors that are important in a given study. Determining how those behaviors should be measured is a potential source of difficulty, especially for beginning researchers. The first step in selecting a measurement system is defining the behaviors of interest, called *target behaviors.*

Target behaviors can be defined either functionally or topographically. *Functional definitions,* which are often used in basic research, specify the consequences of a response; they emphasize the environmental changes that a response produces. Lever presses, key pecks, and button presses are examples of responses defined by their consequences. These responses involve moving an operandum (lever, key, or button) in a particular direction with sufficient force to operate a switch. Each operation of the switch counts as one response. With such arrangements, definition of a response is, in part, a specification of the apparatus used to record it. Such specifications are inherent in the following definition, from a study in which rats emitted lever presses that required a minimum of either 25 g or 200 g force (Makhay, Alling, & Poling, 1994, p. 512):

> A rheostatically controlled electromagnet allowed the minimum force requirement for operation of the left lever [the location and characteristics of which had been specified previously] to be adjusted from 25 to 200 g. With this arrangement, the lever had to be pressed with a force greater than the specified minimum to initiate movement, which proceeded through a downward arc of approximately 0.2 cm so long as at least 25 g of pressure was applied. At the end of this arc, microswitch operation terminated lever movement and a response was recorded.

In this study, lever pressing was reinforced with food under a fixed-ratio 15 schedule. This reinforcement occurred regardless of whether the rat depressed the lever with, for instance, its left forepaw or its nose. These responses were not distinguished by the experimenters and, if they occurred, would have constituted members of the same *operant response class,* which includes all forms of behavior that produce the same kind of reinforcers.

Although studies of functionally defined operant response classes have yielded a great deal of useful information, form as well as function is an important characteristic of behavior. For example, a second-grade student could attract a teacher's attention by either breaking wind or raising a hand, but there is reason to differentiate the two actions. *Topographical definitions* specify the form (physical features) of a response; they emphasize the movements that the response comprises. Such definitions are used in the majority of applied studies and are also common in basic research, in which nonarbitrary response units are of interest. For instance, researchers interested in mating behaviors characteristically define particular response units topographically, not functionally (e.g., Domjan, O'Vary, & Greene, 1988). Table 4.1 provides examples of topographically defined behaviors.

It is generally recognized that good definitions, regardless of whether they emphasize the form or function of behavior, are objective, clear, and complete (Hawkins & Dobes, 1977). As Hartmann (1984, p. 112) explains:

TABLE 4.1. Examples of Topographically Defined Responses

Behavior	Definition	Source
Consummatory or copulatory sexual behavior by male Japanese quail	… consists of the male grabbing in its beak the back of the female's head or neck, mounting the female's back, and then arching its back so as to bring its cloaca in contact with the female. We measured consummatory sexual behavior by recording occurrences of each of these sexual contact responses.	Domjan et al. (1988, p. 506)
Verbal offers of assistance by youths with autism	In order to be scored as correct verbally, the participant had to say "Can I help you?" without prompting within 5 s of the confederate's discriminative stimulus for assistance.	Harris, Handleman, & Allesandri (1990, p. 300)
Self-injurious responses by subjects with developmental disabilities	… were defined as follows: *arm or hand biting*—closure of upper and lower teeth on any portion of the skin extending from fingers to elbow; *face hitting*—audible contact of an open or closed hand against the face or head; and *head banging*—audible contact of any portion of the head against a stationary object (e.g., desk, floor, wall).	Iwata, Pace, Kalsher, Cowdery, Edwards, & Cataldo (1990, p. 13)

Objective definitions refer only to observable characteristics of the target behavior; they avoid references to intent, internal states, and other private events.
Clear definitions are unambiguous, easily understood, and readily para-
phrased. A complete definition includes the boundaries of the behavior, so that
an observer can discriminate it from other, related behaviors. Complete definitions include the following components (Hawkins, 1982): a descriptive name; a
general definition, as in a dictionary; an elaboration that describes the critical
parts of the behavior; typical examples of the behavior; and questionable
instances—borderline or difficult examples of both occurrences and nonoccur-
rences of the behavior.

When behavior is actually measured, verbal descriptions are trans-
lated into operational definitions. *Operational definitions* define particular
responses in terms of how they are measured. In measuring particular
behaviors, either humans or machines transduce stimuli arising from a
subject's actions into numerical data according to specified rules. These
rules are what is stated in an operational definition. They specify not only
the general category of behavior that is of interest, for instance, lever
pressing or self-mutilation, but also the dimensions of behavior that are
important.

Dimensions of Behavior

Dimensions of behavior are specific aspects of defined responses that
can be quantified; they include intensity, frequency, rate, interresponse
time, duration, latency, and accuracy. Choice and permanent products of
responding are also measured in some studies.

Intensity (or force) indexes the vigor with which a particular behavior
is performed. Intensity can be directly indexed in physical units, as when
the downward pressure that a rat exerts on a lever is measured via a strain
gauge. It can also be rated by observers, as when parents score their child's
tantrums on a 5-point scale ranging from very mild (1) to extremely
uncontrolled (5).

Frequency refers to the number of times a response occurs during an
observational period; *rate* specifies the number of times a response occurs
during a defined time period. Rate is expressed as units response per unit
time (e.g., responses per minute) and is calculated by dividing frequency
by time. If, for example, a worker cursed 174 times during an 8-hr observational period, the rate of cursing during that period would be 21.75 curses
per hour (174 ÷ 8). Response rate is determined by *interresponse time* (IRT),
which is the time elapsed between two consecutive responses. Measuring
IRTs has proven especially useful in basic research concerned with the
temporal pattern of responding.

Duration specifies the time that elapses from the beginning to the end of a behavior. Another time-based measure, *latency*, designates the time that passes from the onset of a stimulus to the onset of a response. Reaction time is a latency measure.

Accuracy specifies the extent to which a given behavior is appropriate to the current stimulus conditions. For example, "George Washington" is an accurate (correct) response to the question "Who was the first president of the United States?" But it is an inaccurate answer to the question "Who played center for the Celtics in 1984?"

Choice, which indexes how an organism allocates its behavior (or time) between alternatives, is a common measure in basic research. Choice is often indexed as the percentage (or proportion) of responses allocated to one alternative (i.e., the left lever of a two-lever chamber) relative to total responses. If, for instance, a rat tested in a discrete-trial procedure pressed the left lever of a two-lever chamber 8 times and the right lever 2 times, it chose the left lever on 80% of the trials.

In some settings, permanent products provide a useful index of performance. *Permanent products* are tangible items or environmental effects that result from the occurrence of a particular behavior. If one measures the number of bricks laid by a mason during a workday, one is using a permanent product measure. Like actual behaviors, permanent products can be quantified in terms of rate and accuracy.

In basic research, one is often concerned with the general effects of an independent variable on an operant response class. In such cases, it is reasonable to measure multiple response dimensions. For example, Picker and Poling (1982) were interested in how the proportion of key light illuminations that were followed by food influenced autoshaped key pecking by pigeons. They conducted two experiments. In one, brief illuminations of each of three response keys were followed by response-independent food on 0%, 50%, or 100% of the trials, depending on the color of illumination. Each of the three colors was presented alone during training, and they were presented simultaneously during choice tests. All pigeons consistently directed their initial choice responses and the majority of subsequent responses to the stimulus that was always followed by food, even though during training sessions the response rates of most birds were highest to the stimulus that was followed by food on 50% of the trials.

In a second experiment, different durations of food delivery (e.g. 2 and 10 sec) followed different colors of key illumination. Rate, latency, and percentage of trials with at least one response did not change appreciably as a function of duration of food presentation. However, choice responding was lawfully affected by duration of food presentation, with most responses directed to the stimulus (key color) correlated with the longer food delivery. In this study, the effects of the independent variable depended on the dimension of behavior examined, and choice appeared to be the most sensitive measure. Useful measures of behavior are sensitive, and in that regard, choice was superior to the other measures taken by Picker and Poling (1982). But no single measure is superior in all contexts.

In general, outcome measures in basic research are useful to the extent that they provide meaningful information about behavior and its controlling variables in a particular experimental arrangement. Probably the best strategy for beginning researchers to use in selecting outcome measures is to consult prior investigations in their area of interest. Although they are not sacred, functionally defined operants, such as key pecking, lever pressing, and button pressing, measured along the dimensions of rate and choice, have proven especially useful as a general index of the effects of a range of environmental variables.

In applied studies, emphasize the behaviors and dimensions of behavior constituting the problem that treatment is intended to remedy. The behaviors and behavioral dimensions that constitute a problem in an applied study are often readily apparent and easily stated, as when a child engages in self-injurious face-slapping. In other cases, *social validation procedures* (Kazdin, 1977; Wolf, 1978) may be used to select the aspects of behavior that merit change. One such procedure, termed *subjective evaluation*, involves asking people who will be affected by the intervention (e.g., participants, parents, spouses) what they want changed. For example, parents who seek help in dealing with their child's crying may express a desire to have the behavior decreased in both frequency and duration. A second procedure, termed *social comparison*, involves comparing a participant's behavior with that of peers within given social contexts. The behavior of peers is not deemed changeworthy; hence, differences between what they do and what the participant does appear to constitute the problem for which treatment is sought. Social validation procedures can play a valuable role in evaluating the goals, procedures, and outcomes of applied studies. They were discussed in further detail in Chapter 3.

Especially in laboratory studies, in which collecting and storing data are usually done automatically, it is tempting to measure many dimensions of behavior. There is nothing intrinsically wrong with doing so, if the researcher has determined beforehand why these dimensions are worth measuring, how they will be analyzed as the study proceeds, and which of them will guide decisions concerning the conduct and interpretation of the study. Multiple dependent variables may be necessary to profile fully the effects of an independent variable and can be invaluable in doing so. If, however, the researcher simply collects data on every conceivable dimension of behavior in the interest of finding *something* of importance, the end result is likely to be a pile of meaningless numbers to which there is only one meaninful response: This is garbage—trash it. *Beware the "more is better" trap, and consider how data will be analyzed before collecting them.*

Selecting a Measurement System

In operationally defining the behaviors of interest, a researcher selects a system for quantifying these behaviors along designated dimensions. Behavior can be quantified in many ways, and a great deal has been written concerning measurement strategies (e.g., Bellack & Hersen, 1988; Johnston & Pennypacker, 1993a,b). Good measurement systems are accurate, valid, and reliable, and these concepts merit discussion before we move on to an overview of specific measurement systems.

Accuracy

A measurement system is *accurate* if the value that it yields reflects the true value of the behavioral dimension under consideration. Any difference between the true value and the obtained value constitutes *measurement error*, and the accuracy of measurement decreases as measurement error increases. Simple enough, but what is the "true value" of a variable, and how can that value be obtained?

In practice, the "true value" is the value obtained through the use of the best measurement system available. As Johnston and Pennypacker (1993b, p. 144) indicate:

> There is no single way to obtain true values. Instead, there are only the requirements that (a) the method by which they are obtained must be different from that used to collect the data that are being evaluated, and (b) this method must incorporate extraordinary steps that clearly avoid or remove possible sources of error.

Unless one is using a mechanized recording system of accepted accuracy (e.g., commercial response keys to record key pecks by pigeons), it is good practice to test any measurement system against the best system available. Such tests should be arranged at the onset of experimentation and at least once during each subsequent phase. If at any point the results differ substantially enough that obtained results provide an unsatisfactory reflection of "true values," the everyday measurement system should be calibrated so that the disparity between actual and obtained values is reduced to an acceptable level.

For instance, if self-injurious behavior (SIB) is the target response, frequency of occurrence the dimension of concern, and direct observation the measurement strategy, a videotape of the subject(s) should be secured at the beginning of the study and during each phase thereafter. Experts, including the directors of the study, carefully score the tape, perhaps played in slow motion, to determine what they consider to be the "true" frequency of occurrence of SIB. Then, the same tape is scored by the actual observers. If the measures differ substantially, calibration of the everyday measurement system is required. When humans are used to record behavior, calibration usually involves altering response definitions, providing additional training, or changing the conditions of observation.

Validity

In a general sense, a measurement system is *valid* to the extent that it measures what it purports to measure. When indirect measures of behavior are used, such as intelligence tests or personality inventories, the issue of validity is complex and vexing, for there is no agreement as to what, exactly, such instruments are intended to measure. In fact, there is no "thing" that they measure; thinking of "intelligence" or "personality" as a physical entity is an error of reification with unfortunate consequences (Gould, 1981).

In attempting to evaluate the adequacy of devices that measure nothing with a discrete physical existence, psychometricians have distinguished several subtypes of validity—including construct, content, concurrent, predictive, criterion, face, and convergent (Kazdin, 1992)—and have written thousands of pages discussing how these subtypes can be distinguished and measured. In summary, validity in its various aspects is assessed by having experts evaluate a measurement system and by determining whether the scores yielded by that system allow for prediction of performance in other situations. Unfortunately, experts often disagree, and the success of prediction varies with the criterion measure. For these reasons, there is continuing controversy over the validity of most indirect assessment devices.

Indirect assessment is frequently used through necessity, as when some hypothetical construct such as masculinity or femininity is of concern. But it can also be used for convenience in indexing actual behavior, as when a researcher has a person complete a questionnaire regarding his or her physical activity. In such cases, it may be possible to collect direct measures of behavior that can be compared to the indirect measures. For example, if a person reports working out at a particular gym three days a week for an hour each time, it would be easy enough to have an observer determine the truth of the assertion. Checking sign-in sheets at the gym would provide another check on accuracy, albeit a less direct one. If data collected through direct observation parallel self-reports, the accuracy of the latter is documented.

Reliability

In general, an assessment system is *reliable* to the extent that it yields a consistent outcome so long as the actual behavior of concern does not change. Salvia and Ysselkyke (1981) emphasize the worthlessness of an unreliable instrument by comparing it to an easily stretched rubber ruler to be used to measure peoples' heights. Such a ruler would not produce consistent measurements even if a single person were repeatedly measured under the same conditions because the value obtained would depend primarily on how much the ruler was stretched, not on the person's physical dimensions. To extend the ruler analogy further, a rigid ruler could conceivably be a reliable measure of a person's weight by, say, measuring the person's height and multiplying it by a fixed number of pounds per inch, but it could not be valid. If used consistently, a rigid ruler would yield repeatable data and hence be reliable, but the dimension assessed would not be weight as it is usually conceived. Therefore, the measure would be worthless.

Describing Measurement Systems

Measurement systems can be conveniently classified and analyzed in terms of four general characteristics: (1) whether humans or machines serve as transducers, (2) whether behavior is measured directly or indirectly, (3) whether behavior is measured continuously or intermittently, and (4) whether behavior is measured in the actual situation of concern.

Humans vs. Machines

Most basic research studies use machines to convert the subject's actions into numbers; most applied studies use human observers for the same purpose. Neither humans nor machines are clearly superior. Both have strengths and weaknesses that make them especially useful in particular applications. Three potential advantages of mechanized recording are:

1. Highly accurate recording over long time periods. Well-calibrated mechanized systems can produce exceptionally accurate and reliable measures of behavior. For example, a computerized system that measures each switch closure produced by an undergraduate's pressing of a button can

produce a highly accurate record of this behavior over extended time periods. Human observers faced with the same task would probably do less well.

2. Inexpensive data collection. Although data-recording devices may be expensive to purchase, they can be used relatively cheaply for many years. Human observers, in contrast, are costly if used often.

3. Nonreactive recording. *Reactivity* refers to the effects of an assessment procedure on the behavior being assessed. In some cases, the mere presence of a human observer disrupts a subject's responding, whereas an automated recording device does not. For instance, audiotaping a family's discussions at dinner is less likely to alter content than is having an actual observer present. The audiotapes would eventually be scored by humans; thus, the data collection procedure would involve both a machine and a human observer. Of course, not all automated recording systems are nonreactive. Placing a pressure-sensitive band (plethysmograph) around the penis to index male sexual arousal, for instance, may deflate what it is emplaced to measure.

The major limitation of mechanized recording is that there are many important behaviors that machines cannot readily score. Human observers, if appropriately trained, can provide a relatively accurate index of a wide range of responses. This is their main advantage and the reason they are used in most applied studies. There are, however, several unique problems associated with human observers. One is that the manner in which they apply response definitions sometimes shifts over time, a phenomenon known as *observer drift*. Another is that an observer may be biased. In general, reported data are *biased* if they consistently differ from true values in a particular direction. A biased observer either consistently overestimates or consistently underestimates target behaviors.

Observer expectations about the outcome of an intervention may lead to biased evaluations only during a particular phase of an experiment; such an inconsistent bias is a greater problem than a consistent one. Consider, for instance, a teacher who is confident that methylphenidate (Ritalin) improves the overall deportment of children diagnosed as hyperactive. If asked to rate hyperactivity in a student under conditions in which Ritalin was and was not given, that teacher is likely to rate the child as improved under the former condition, even if the child's actual behavior did not change. Such an outcome would lead to a faulty conclusion about the drug's effectiveness.

To reduce the probability of such problems, it is characteristic to employ so-called *blind* observers, who are not informed regarding experimental conditions or expected changes in behavior. In addition, carefully constructed observational systems usually are used; these systems direct observers' attention to actual behaviors of importance. Finally, calculating interobserver agreement, which is done by comparing the data recorded by two people who simultaneously and independently monitor the same subject, provides a check on the general adequacy of observation. The calculation of interobserver agreement is of sufficient importance in applied behavior analysis that a separate section is devoted to it later in this chapter.

Direct vs. Indirect Measures

As discussed in the section on validity, *direct measures of behavior are superior to indirect measures and therefore should be used whenever possible.* Of course, if a researcher is interested in private events, which are stimuli and responses that occur within a subject's own skin (e.g., thoughts, feelings), there is at present no possibility of measuring them directly. Private events characteristically are quantified through self-reports (discussed below), which are perhaps best viewed as inevitably suspect but sometimes unavoidable.

Continuous vs. Intermittent Measurement

In general, procedures that provide continuous measures of behavior are superior to those that provide occasional measures. The reason is simple: If behavior is not continuously monitored, one must perforce assume that behavior during monitored periods of time provides an accurate reflection of performance during unmonitored periods, and this assumption may not be valid. Intermittent monitoring, which is frequently termed *time-sampling* in the applied literature, is especially troublesome with behaviors that occur infrequently and are therefore likely to be "missed."

In most basic research, experimental sessions are relatively short (0.5–2 hr), and it is possible to arrange continuous measurement of target behaviors. In some applied studies, however, much larger samples of behavior are of interest, and practical or ethical considerations may render continuous observation difficult or impossible. For instance, a researcher interested in the effects of a token economy on the behavior of psychiatric inpatients would in all likelihood be concerned with their activities throughout the day and night, but probably would not be able to monitor them directly and continuously around the clock. In such a case, continuous monitoring might be arranged for certain periods of the day, for example, times when problems characteristically occur, such as during meals. During the rest of the day, behavior would be monitored occasionally, for instance, by observing for a brief period on average every 30 min. *When continuous monitoring is not possible, the more frequently behavior can be monitored, the better.*

Naturalistic vs. Contrived Measurement

A particular behavior can be monitored either in the situation in which it naturally occurs (naturalistically) or in some other setting, such as a laboratory or clinic. For example, a researcher interested in the effects of sleep deprivation on driving an automobile could examine how that variable affected the performance of actual drivers on a roadway or in a setting that simulated actual driving. If the latter tack were taken, assessment would be accomplished via *analog methods*. Analog methods simulate the situation of concern in a way that allows behavior to be easily and safely monitored. For example, a driving simulator requires the operator to engage in behaviors similar to those needed to drive an automobile on the highway. Measuring how behaviors in the simulator are affected by sleep deprivation would allow one to examine safely and easily whether that variable is likely to reduce peoples' ability to operate a motor vehicle safely.

One important potential problem with analog methods is low *ecological validity*: The effects observed in the analog situation may not accurately reflect a subject's performance in the actual situation of concern. For example, after training in "safer sex" techniques, a person tested in a role-played laboratory simulation may consistently indicate that condom use is a prerequisite for sexual activity with a new partner, but behave very differently when a real and desirable lover is available. When such disparities do not occur and ecological validity is not a problem, analog assessments are useful and convenient strategies for monitoring behavior.

Techniques for Measuring Behavior

Many techniques are available for quantifying behavior, including standardized personality and intelligence tests, checklists and rating scales, mechanical recording, self-reports, and direct observation. Standardized personality and intelligence tests are for the most part of little use in behavior analytic research, for they are indirect measures of questionable validity and sensitivity. Although checklists and rating scales also are indirect measures, they are easy and inexpensive to use and occasionally prove useful as rough-and-ready indices of intervention effects in applied settings. The critical requirement when these devices are employed is correspondence between raters' evaluations and important aspects of the subject's behavior. When it is clear that checklists or rating scales provide accurate measures of target behaviors, they are simple and cost-effective assessment devices. One area in which checklists and rating scales have been used to good advantage is in evaluating side effects of psychotropic medications (Poling, Gadow, & Cleary, 1991). Such devices allow simultaneous monitoring of far more aspects of a subject's physical status and behavior than could be practically monitored through direct observation.

As previously discussed, mechanical recording devices have long played a valuable role in basic research. Two devices, the event recorder and the cumulative recorder, produce a lasting visual record of responding (and other significant events) and deserve special consideration.

Event Records and Cumulative Records

An *event record* is made by a device in which one or more marking pens are mounted in such a way that they trace a line on a horizontal strip of paper that is pulled under them at a constant speed. Each pen can move side to side on the strip from one position to another, producing a continuous graph as a function of inputs from an observer or an automated device. The position of the pen corresponds to whether or not a particular event has occurred. By convention, the pen moves to the up position when an event or activity begins, stays there so long as the event occurs, and returns to the down position when the event ends. If, for example, an observer wanted to produce an event record of a dog's scratching itself, an event recorder could be programmed so that pressing a red button moved the pen up and pushing a white button moved the pen down. If an observer pressed the two buttons at the appropriate times, accurate and lasting visual records of the behavior would be obtained.

With the addition of more pens, each corresponding to a different aspect of the environment, an event record can provide a good deal of information. Figure 4.1 shows an event record depicting the performance of a hypothetical human subject responding under a multiple fixed-ratio 5 (FR 5) extinction schedule. The reinforcer is delivery of a nickel; the light is on when the FR 5 schedule is in effect, off when the schedule is extinction.

Event records are very useful for portraying temporal relations between stimuli and responses. But their utility is limited to collecting data over relatively short periods of time. In order to prevent loss of information, the paper must move fast enough to provide some white space between the most rapidly occurring responses. With many experimental preparations, response rates as high as 5 responses per second can occur, so the paper speed should be at least 5 millimeters per second. At this speed, a 30-minute session would be represented by a strip of paper 9

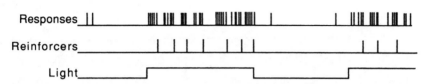

<figure>FIGURE 4.1.   An event record depicting the performance of a hypothetical human subject respondng under a multiple fixed-ratio 5 (FR 5) extinction schedule. The light is on when the FR 5 component is in effect and off during the extinction component.</figure>

meters long. Such a large and complex stimulus rarely clarifies relations between independent and dependent variables. For this reason, event records usually are used as occasional, short-duration supplements to other data collection procedures.

A *cumulative record* is similar to an event record, with the exception that each response moves the pen a fixed vertical distance, called the "step size." After a preestablished total of such vertical steps has been traversed, the pen resets to the initial level. In addition to moving upward, the pen on must cumulative recorders can also be made to move downward and to the right, making a short diagonal slash. Such slashes conventionally are used to designate the delivery of a reinforcer. Most cumulative recorders include a single-pen event recorder that is used to indicate changes in stimulus conditions or other important events.

When the step size is small and the paper speed is slow, the cumulative record can portray in a small space much information about rates and temporal pattern of responding over time and in relation to specific stimulus changes. For example, Figure 4.2 shows cumulative records from a study in which rats responded under a concurrent schedule of food and ethanol (beverage alcohol) reinforcement (Poling & Thompson, 1977). Food was always initially available under a fixed-interval (FI) 26-sec schedule, and tap water containing 8% ethanol was always available under an FR 1 schedule. Across conditions, each response that produced ethanol also increased the FI food schedule by a specified interval. The cumulative records in Figure 4.2 show that a 32-sec delay decreased overall rate of ethanol-maintained responding relative to baseline (0-sec delay) levels, but did not affect the temporal patterning of responding. In contrast, a 512-sec delay shifted the temporal pattern, so that all ethanol-maintained responding occurred late in the 1-hr session. By responding in this way, the rats received roughly 75% of the potentially available food and also received almost as much ethanol as in the baseline condition. Had they responded to produce ethanol earlier in the session, far less food

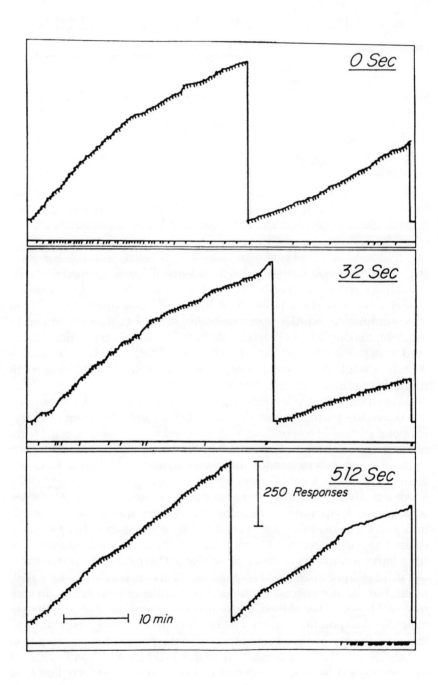

would have been available. The cumulative records depict this outcome more easily and clearly than does a verbal description.

The popularity of cumulative records as devices for depicting results in published studies has diminished greatly from the early years of behavior analysis (Poling, 1979). Nonetheless, cumulative records provide an exceptionally precise depiction of moment-to-moment changes in behavior and are still used by many basic researchers to keep track of how their subjects are behaving. *Anyone conducting basic research is wise to collect and respond to cumulative records on a daily basis, for doing so is unparalleled in bringing the behavior of the researcher under the control of the appropriate subject matter.*

Self-Reports

Self-reports require a subject to collect data on his or her own behavior. They are essential dependent variables if one is interested in stimuli and responses directly accessible only to the behaving person. Such private events are not automatically excluded from a behavioral analysis. As Skinner (1974, p. 16) pointed out:

> [Radical behaviorism] does not insist on truth by agreement and can therefore consider events taking place in the private world within the skin. It does not call these events unobservable and does not dismiss them as subjective. It simply questions the nature of the object observed and the reliability of the observations.

Unless there are public accompaniments of private events, it is virtually impossible to ascertain the reliability, accuracy, and validity of self-reports of those events. This impossibility constitutes a serious problem for one committed to sound research methodology.

Nonetheless, self-reports have a role to play in behavior analysis. Self-reports are not intrinsically inaccurate, and they often provide the only means of indexing private events of theoretical significance (e.g., covert rules). Moreover, self-reported internal states often constitute a significant part of a clinical problem, as in the dysphoria associated with depression,

---

←

FIGURE 4.2.   Representative cumulative records of lever pressing by a rat during conditions in which lever presses that produced access to an 8% ethanol solution did not delay food delivery, delayed food delivery by 32 sec, or delayed food delivery by 512 sec. The upper track of each record was stepped by presses on the lever that produced food; food presentation is indicated by a diagonal pip. Ethanol-reinforced lever presses are indicated by diagonal slashes on the bottom horizontal track. Food initially was available under a fixed-interval schedule of food delivery. From Poling and Thompson (1977). Copyright 1977 by the Society for the Experimental Analysis of Behavior. Reproduced by permission.

and ignoring such states can lead to an incomplete analysis of treatment effects. Finally, self-reports provide a convenient way to detect adverse side effects of treatments that are not easy to predict or otherwise measure.

Direct Observation

Direct observation involves someone actually watching the subject and recording the subject's behavior. Many different observational systems are available; some of them are quite complex and allow for the simultaneous quantification of several behaviors. For example, Marholin, Touchette, and Stewart (1979) used direct observation to examine the effects of chlorpromazine (Thorazine) on several behaviors of mentally retarded adults. Among the responses measured were compliance with verbal requests, accuracy and rate of performance on workshop tasks, time on task, eye contact, talking to self, talking to others, standing, walking, being within three feet of others, approaching others, and touching others. Quantifying all these behaviors allowed for a broad-spectrum assessment of the behavioral effects of the drug, which did not prove beneficial for any participant.

Figure 4.3 shows how a single hypothetical response (self-stimulatory behavior) would be quantified under two different observational systems during one 10-min observational period during baseline and treatment. This figure presents hypothetical data for frequency and duration measures obtained through real-time observations. It also presents percentage occurrence measures.

In *real-time observations*, the onset and offset of the target behavior is recorded as it occurs in time. Frequency and duration measures are easily obtained from such observations, and real-time observations have been characterized as powerful, rigorous, and flexible (Hartmann & Wood, 1982). But real-time observations are difficult to accomplish unless human observers are aided substantially by automated devices (e.g., computers).

Percentage occurrence measures characteristically are collected through the use of interval recording procedures. In an *interval recording* procedure, the period during which behavior is to be observed is divided into discrete, and relatively brief, observational intervals. For example, in Figure 4.3, the 10-min period of interest (perhaps the first 10 min of lunch) was divided into 1-min observational intervals. Observers used a data sheet (shown in the figure) to record whether the behavior of interest occurred in each of these intervals. A "+" indicates that the behavior occurred in a particular interval, whereas a "−" indicates that the behavior failed to occur. Intervals with a "+" (the behavior occurred) are termed "scored intervals" (or "occurrence intervals"), and intervals with a "−" (the behav-

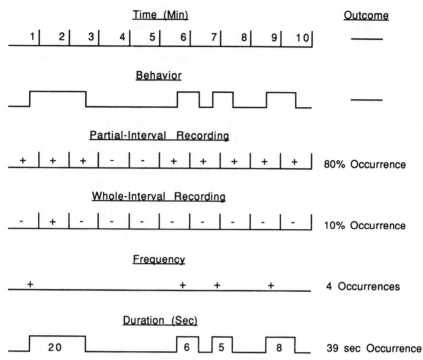

FIGURE 4.3.   Frequency, duration, and time-sampling measures of a hypothetical behavior.

ior did not occur) are termed "unscored intervals" (or "nonoccurrence intervals").

Most often, an interval is scored if the behavior of interest occurred at any point during that interval. In some cases, however, an interval is scored only if the behavior is occurring when the observational interval begins and continues throughout the observational period. The former scoring system is termed *partial-interval recording*; the latter, *whole-interval recording*. Values obtained with whole-interval recording can never exceed those obtained with partial-interval recording. Whole-interval recording is of value when the target behavior must occur without interruption to be appropriate. Figure 4.3 presents data for both partial-interval and whole-interval recording.

Although interval recording is popular in applied behavior analysis, the technique has several limitations (Hartmann & Wood, 1982). The method provides a good index of the duration of behavior only when observational intervals are very brief. Moreover, the method may over- or underestimate the frequency of behavior, depending on the duration of intervals. If intervals are long, the response may occur more than once, yet only one response will be recorded. But if intervals are short, the same response may be recorded in two or more intervals. For these reasons and others, researchers should not blindly follow the convention of employing interval recording in applied studies. Nor should they assume that the percentage of observational intervals in which the behavior occurred (percentage occurrence) is tantamount to either frequency or duration.

There are a number of different observational systems in addition to those introduced above (see Bates and Hanson, 1983; Hartmann & Wood, 1982). As Figure 4.3 illustrates, the various systems do not necessarily yield comparable outcomes. Therefore, the person responsible for designing the observational system should be aware of the range of procedures commonly employed and their characteristics and should select one that is likely to provide an accurate reflection of target behaviors in the situation of concern.

Calculating Interobserver Agreement

No matter what the procedure employed, the accuracy of humans in recording behavior inevitably is open to question. In response to the potential problem of inaccurate observation, it has become standard practice for applied researchers to calculate and report a measure of interobserver agreement (Page & Iwata, 1986). Interestingly, interobserver agreement is rarely reported when humans are used to quantify the behavior of nonhumans in basic research (Poling, 1985; Poling, Monaghan, & Cleary, 1980). There is no good reason for this disparity, and *it is good practice to calculate interobserver agreement whenever humans are used as transducers, regardless of the species being observed.* A good general rule is to ensure that interobserver agreement is calculated during each phase of a study and for at least 25% of the total observations.

Measures of interobserver agreement (IOA) summarize the extent to which independent observers agree on the occurrence and nonoccurrence of specified behaviors. Calculating IOA enhances the believability of observations through consensual validation, which is well accepted in science generally (Kuhn, 1970). The essence of *consensual validation* is the reproducibility of observations: If two (or more) independent observers consistently can agree as to whether or not a phenomenon has occurred, that phenomenon is consensually validated, and others have reason to assume that the phenomenon is a real event, adequately defined, and measured without substantial bias under conditions allowing for reasonably accurate assessment (Hawkins & Fabry, 1979; Johnson & Bolstad, 1973). It is important to recognize, however, that the mere fact that two observers agree does not *prove* that they are accurately monitoring behavior. The best way to evaluate accuracy is to compare obtained scores to true scores, as discussed previously. Demonstrating that an observational system yields high IOA is not a substitute for calibrating that system. Nonetheless, observational systems with low IOA are never acceptable, and the calculation of IOA is an established methodological convention.

There are several techniques for calculating IOA, and they do not necessarily produce comparable results.* During the late 1970s, techniques for calculating and reporting IOA were discussed at considerable length (e.g., Hawkins & Dotson, 1975; Hawkins & Fabry, 1979; Kratochwill & Wetzel, 1977; Repp, Deitz, Boles, Deitz, & Repp, 1976). According to Page and Iwata (1986), the most commonly reported indices of agreement are total, product–moment correlation, interval, occurrence, nonoccurrence, and, more recently, kappa.

Total Agreement

Total (or frequency ratio) agreement is determined for a given observational session by calculating the number of occurrences of the target behavior recorded by each observer, then dividing the smaller number by the larger and multiplying by 100. If, for instance, Observer A recorded 50 instances of cursing by a child at lunch and Observer B recorded 55 instances, total agreement for that observational period would be $50/55 \times 100$, or 91%. Although total agreement is easy to calculate, there is no assurance that the two observers ever agreed on the same occurrences of the target behavior. This uncertainty is a significant weakness of the procedure.

Product–Moment Correlation

This measure, which can range from $+1.0$ to -1.0, expresses the extent to which the scores recorded by two observers covary across a number of observational periods. It is calculated according to the following formula:

---

*There is no agreement on the terms that should be used to describe the different methods for calculating IOA. We use the terms employed by Page and Iwata (1986), but the reader should recognize that other authors may use different terms or the same terms with different meanings.

$$r = \frac{\Sigma\,(d_x)(d_y)}{N\sigma_x\sigma_y}$$

where x represents the scores of one observer and y the scores of the second observer; d_x and d_y are deviations of each score from its mean; N is the number of sessions over which the correlation is computed; and σ_x and σ_y are the standard deviation of the x and y scores, respectively.

The higher the correlation coeffecient, the greater the covariance, which is the extent to which the scores of one observer correspond to those of the second observer, when both sets of scores are rank-ordered. If the rank orderings are in perfect parallel, the correlation coeffecient is +1.0. If they are directly opposite, it is −1.0. A correlation of +1.0 would occur if, for example, Observer A recorded progressively higher scores across five observational periods and Observer B did the same.

Correlation measures of IOA have two major weaknesses. One is that it is possible to obtain high coeffecients when one observer consistently scores more occurrences of behavior than the other observer. A second is that correlation coeffecients provide no assurance that observers agreed on the occurrence of any behavior.

Exact Agreement

Exact agreement can be determined when observers record the frequency of a target behavior within discrete intervals. Agreement is calculated by the following formula:

$$\frac{A_{freq}}{A_{freq} + D} \times 100$$

where A_{freq} is the number of intervals for which the two observers recorded exactly the same number of occurrences of the target behavior and D is the number of intervals for which two observers did not record exactly the same number of occurrences of the target behavior.

Exact agreement is more sensitive to agreement on specific instances of behavior than are other common indices, which is a point in its favor. In some cases, however, the method is so conservative that it may be difficult or impossible to demonstrate reasonable levels of IOA (Page & Iwata, 1986).

Interval Agreement

When interval recording methods are used to collect data, IOA is often expressed as the percentage of total intervals in which the ratings of the two observers agreed. This measure of IOA is calculated according to the following formula:

$$\frac{A}{A + D} \times 100$$

where A is the number of intervals for which ratings of the two observers agreed and D is the number of intervals for which the ratings of the two observers disagreed.

Although this formula historically has been the standard method for calculating IOA, it may provide misleading information when the target behavior occurs either very frequently or very infrequently. Consider a situation in which two observers are independently and randomly marking interval data sheets with "+" and "−" signs. Their markings would agree due to chance alone on some proportion of the intervals, and the extent to which this chance agreement occurred would vary according to their independent rates of recording "+" and "−" signs. If, for example, both observers always recorded a "+," their agreement would be 100%. If, however, both recorded a "+" for half the intervals and a "−" for half, their agreement over many trials should be 50%. This measure, which represents "chance" agreement, is readily calculated:

$$\frac{(O_A \times O_B) + (N_A \times N_B)}{T^2} \times 100$$

where O_A and O_B represent the proportion of trials in which observers A and B indicated that the behavior occurred, respectively; N_A and N_B represent the proportion of trials in which observers A and B indicated that the behavior did not occur, respectively; and T represents the total number of observational intervals.

When a target behavior occurs very frequently, overall IOA can be high even though the observers rarely agree on the occurrence of the behavior. Similarly, when a target behavior rarely occurs, overall IOA can be high even though the observers rarely agree on the nonoccurrence of the behavior. In recognition of this problem, researchers have suggested that it is reasonable to calculate and report measures of IOA separately for intervals in which one or both observers indicate that the behavior did occur (scored-interval, or occurrence, agreement) and for intervals in which one or both observers indicate that the behavior did not occur (unscored-interval, or nonoccurrence, agreement). The following equations are used to calculate occurrence and nonoccurrence agreement, respectively:

$$\frac{A_{occ}}{A_{occ} + D_{occ}} \times 100 \tag{1}$$

$$\frac{A_{non}}{A_{non} + D_{non}} \times 100 \tag{2}$$

In Equation (1), A_{occ} refers to total agreements in intervals for which both observers indicate that the behavior occurred and D_{occ} refers to total disagreements in intervals for which one or both observers indicate that the behavior occurred. In Equation (2), A_{non} refers to total agreements in intervals for which one or both observers indicate that the behavior did not occur and D_{non} refers to total disagreements in intervals for which one or both observers indicate that the behavior did not occur.

Kappa

There are no simple conventions concerning how IOA should be calculated in a given study or what level of IOA is acceptable. There appears to be some support for the notion that IOA should be at least 80% (e.g., Hersen & Barlow, 1976), but this value is arbitrary. Moreover, it may not exceed chance levels. Statistical procedures are available for determining whether obtained levels of IOA exceed chance levels; one such statistic is kappa (Cohen, 1960). Kappa calculates the proportion of agreements, with correction for agreements due to chance. It is calculated according to the following formula:

TABLE 4.2. Twelve Characteristics of Good Measurement Systems

1. They actually measure what they are supposed to measure (i.e., they are valid).
2. They measure behaviors that are relevant to the experimental question.
3. They measure important dimensions of the target behavior.
4. They yield quantitative information.
5. They measure behavior directly.
6. They measure behavior repeatedly over time.
7. They yield measures that precisely reflect the level of behavior or its products (i.e., they are accurately calibrated).
8. They yield consistent measures if behavior does not change (i.e., they are reliable).
9. They do not affect the level of the target behavior (i.e., they are nonreactive).
10. They are acceptable to those whose behavior is being monitored and those who will use the information obtained.
11. They measure behaviors capable of being affected by the intervention (i.e., they are sensitive).
12. They are practical (i.e., compatible with available resources).

$$\frac{(P_o - P_c)}{(1 - P_c)}$$

where P_o is the proportion of obtained agreements and P_c is the proportion of agreements expected due to chance.

Summary and Conclusions

Many different techniques can be used to quantify behavior, and they differ in many ways. Nonetheless, measurement systems that are generally useful share certain characteristics, which are listed in Table 4.2. The presence or absence of some of these characteristics may be obvious on superficial inspection of a proposed or extant system; others (e.g., reliability, accuracy) must be evaluated empirically. All merit consideration, for a study can be no better than the system it uses to measure behavior. In many cases, especially in applied settings, it will require substantial effort to select and define target behaviors and to develop workable procedures for observing and recording those behaviors with accuracy. The effort is justified.

CHAPTER 5

Within-Subject Experimental Designs

There are many ways to arrange conditions across time and subjects and at the same time maintain the fundamental features of within-subject designs. Applied behavior analysts have assigned names to a number of specific arrangements and have considered the strengths and weaknesses of these designs at some length. For example, Baer et al. (1968) indicated in the inaugural issue of the *Journal of Applied Behavior Analysis* that reversal and multiple-baseline designs were generally useful in providing a believable demonstration of the effects of an intervention on specified outcome measures. These designs continue to be used in many applied studies, but several options are available. This chapter discusses common within-subject designs, specifically, case study, A–B, withdrawal,* multiple-baseline, multiple-probe, changing-criterion, alternating-treatments, and concurrent-schedule designs.

Unlike applied researchers, those who conduct basic research rarely refer to designs by name in their articles. They characteristically arrange conditions in a manner analogous to what applied behavior analysts term withdrawal, concurrent-schedule, or probe designs.

Case Study Designs

The case study design, or B design, has a long and controversial history of application in traditional clinical psychology (Barlow & Hersen, 1984; Barlow, Hayes, & Nelson, 1983). In case studies, treatment is introduced prior to or in conjunction with initial observation of the target

---

*Withdrawal designs are also termed "reversal designs," although the latter designation may be misleading (see Barlow & Hersen, 1984).

behavior, which may or may not be precisely defined and quantified. Observed changes in the target behavior are attributed to treatment on the assumption that no such changes would have occurred in the absence of treatment. Because there is no strong empirical support for this assumption, the assumption is tenuous, and the case study design does not support strong conclusions about the relation between independent and dependent variables. This lack of support for conclusions is its primary weakness.

Several strategies have been suggested for improving case study designs. Kazdin (1982), for example, indicates that multiple case studies with continuous assessment of target behaviors provide a reasonable assessment of treatment effectiveness under conditions in which more rigorous assessments are impossible. Even with such refinements, the primary role of case studies in a science of behavior appears to be in generating questions that can subsequently be subjected to rigorous experimental analysis.

A–B Designs

The A–B design involves a baseline (A) phase followed by the introduction of treatment (B). Characteristically, the target behavior is monitored repeatedly during both phases. A treatment effect is demonstrated by showing that performance differs from one phase to the next. Figure 5.1 presents the results from a hypothetical experiment in which an A–B design was employed and the intervention appeared to reduce the rate of occurrence of the target behavior.

The A–B design is better than the B design insofar as the collection of baseline data provides an empirical basis for predicting what the level of the target behavior would have been had no intervention occurred. However, the prediction is always of questionable accuracy. Although past behavior is the best predictor of future behavior, it is an imperfect predictor, and it is possible that changes in behavior observed in the B phase would have occurred regardless of whether treatment was introduced. That is, any changes in behavior observed in the B condition may be the result of an extraneous variable that coincidentally became operative at the time treatment was introduced or near that time.

Consider, for example, a situation in which a researcher posits that providing verbal instructions will increase the on-task behavior of a hyperactive child. On-task behavior is appropriately defined and measured during an initial baseline (A) phase, followed (after behavior in the A phase has stabilized) by the intervention (B). After a few days of exposure

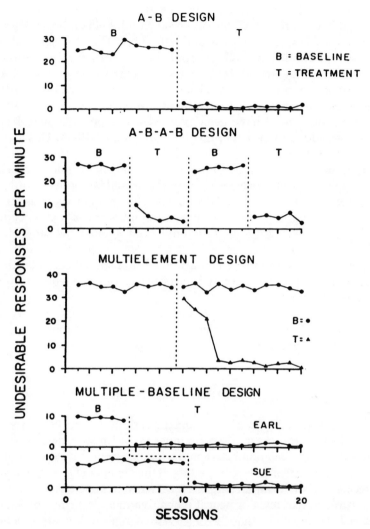

FIGURE 5.1. Hypothetical data from experiments utilizing A–B, A–B–A–B, multielement (or alternating-treatments), and multiple-baseline designs. Each set of data suggests that the treatment was effective in reducing undesired behavior.

to treatment, levels of on-task behavior clearly increase, and remain stable for several days. This finding suggests that the intervention was effective. It is possible, however, that some extraneous variable became operative coincident with the introduction of treatment and was actually responsible for the observed improvement. Perhaps at the onset of the B phase, another child, a favored playmate of the subject child but a potent disrupter of on-task activity, moved away and left school. The playmate's absence, not verbal prompting, actually accounted for the observed increase in on-task behavior. If the researcher failed to detect the confounding effects of the playmate's absence, which is likely, the researcher would attribute the increased on-task behavior to treatment and would err in doing so. This error probably would have been avoided had a posttreatment baseline phase been added to the experiment, making it an A–B–A. This is a rudimentary withdrawal design.

Exposing other subjects to the intervention also would decrease the likelihood that the effects of verbal prompts would be confused with those of a schoolmate's absence, for the effects of such an extraneous variable would be limited in space and time. Although a friend's absence might confound the effects for John, tested in Grand Rapids, that same absence could not similarly affect Janet, tested in Kalamazoo two weeks later. Independent A–B designs, when arranged in a special configuration involving the temporally staggered introduction of treatment, constitute a multiple-baseline design.

Like the case study design, the A–B design supports only weak conclusions and can be recommended only when other, more compelling designs are not tenable.

Withdrawal Designs

Withdrawal designs play a very important role in both basic and applied behavioral research. The manner in which such designs demonstrate a functional relation between the independent and dependent variable is straightforward and compelling: If the dependent behavior changes appreciably from the initial baseline level when treatment is implemented, and returns to that level or near it when treatment is terminated, there is good reason to believe that the observed changes in the target behavior reflect the action of treatment. It is possible, of course, that some extraneous variable coincidentally begins to affect the subject's behavior when treatment is introduced, remains operative throughout the course of treatment, and ceases when treatment ends. Unless the extraneous variable is actually associated with treatment, the likelihood that this will happen is small (but never nonexistent) and grows smaller with each additional implementation and termination of treatment. Data showing the results of a hypothetical experiment employing an A–B–A–B design are presented in Figure 5.1.

Consider our example of the hyperactive child, discussed previously. Had the design been extended to an A–B–A, on-task behavior would have remained high during the second A phase, unless the schoolmate returned, an unlikely occurrence.

Many variants of the withdrawal design are possible (see Barlow & Hersen, 1984). When used appropriately, all of them are capable of supporting firm conclusions concerning functional relations between independent and dependent variables. A study by Marholin et al. (1979), described briefly in Chapter 4, demonstrates the use of a withdrawal design to evaluate an intervention already in effect. These investigators evaluated the effects of a neuroleptic drug, chlorpromazine (Thorazine), in four institutionalized mentally retarded adults. The participants were first observed during 19 days in which chlorpromazine was administered. (A fifth person was also studied under a similar but more complex design, but for simplicity this person will not be considered here.) Several behaviors were carefully measured in workshop and ward settings (e.g., compliance with verbal requests, accuracy and rate of performance of workshop tasks, time on task). The initial treatment phase was followed in sequence by a 23-day drug-free (placebo) period and a 25-day period in which medication was reinstated. The effects of withdrawing chlorpromazine differed somewhat across subjects, but some desirable behaviors did emerge when the drug was withheld (see Figure 5.2). Given this outcome, continued drug treatment of these patients was indefensible.

When appropriately arranged, withdrawal designs can be used to evaluate treatment interactions, as well as to compare multiple interventions in isolation. Two basic considerations apply to the use of all withdrawal designs. One is that conditions should not be changed until behavior is relatively stable over time. In general, behavior is stable if, across successive observations, variability is minimal and there is no upward or downward trend in performance. Stable data are important because they facilitate analysis of the effects of an intervention and also provide a measure of its full (steady-state) effects.

Most behavior analytic research is concerned with the analysis of steady states. Indeed, as Perone (1991, p. 139) indicates:

> [B]y contemporary standards the production of steady states has become an essential feature of experimental method in the analysis of free-operant behavior. Each condition is continued until a steady state is observed, at which point another condition is imposed, and the effects of the experimental manipulations are evaluated by comparing the steady-state performances from the last few sessions of the various conditions.

Perone (1991, p. 139) goes on to explain that researchers interested in analyzing steady-state behavior must overcome three obstacles:

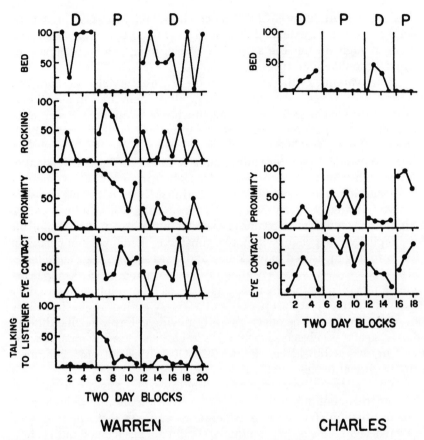

FIGURE 5.2.   Percentage of observational intervals in which a variety of social behaviors were emitted by people with mental retardation during drug (D) and placebo (P) conditions. From D. Marholin, P. E. Touchette, and R. M. Stuart, "Withdrawal of chronic chlorpromazine: An experimental analysis," *Journal of Applied Behavior Analysis*, Vol. 12, p. 165. Copyright 1979 by the Society for the Experimental Analysis of Behavior. Reprinted by permission.

> First, they must exert enough control over experimental and extraneous vari-
> ables to engender a steady state. Second, they must impose such control long
> enough for the steady state to occur. Finally, they must recognize the steady
> state when it does occur.

Recognition of the steady state can be based on visual inspection of obtained data or on quantitative analysis of those data. The rules used to determine when behavior is stable constitute *stability criteria*. Stability criteria can vary from minimal to quite rigorous. For example, a criterion of 5 consecutive observations (experimental sessions) with no visually evident trend in rate of responding is less demanding, and more easily met, than a criterion of 5 consecutive sessions in which the average rate of responding in each individual session varies by no more than 5% from the average rate of responding across all the sessions.

Stability criteria play a critical role in how most within-subject experiments are conducted, insofar as conditions often are changed only when behavior stabilizes. Stability criteria that are too lenient may lead to premature condition changes and thereby prevent the researcher from detecting treatment effects that would be apparent with longer exposure. Unduly strict criteria may never be met.

Actual experience with the characteristics of behavior under conditions similar to those of interest in a given experiment is invaluable in establishing good stability criteria; familiarity with the criteria used by others working in similar areas is also useful. As a rule, good stability criteria meet these requirements:

1. They can be described in objective terms.
2. They can be met within a reasonable period of time.
3. They are similar to stability criteria used by other researchers.

Practical as well as theoretical considerations influence the choice of stability criteria in applied research, and practical exigencies may dictate that conditions be changed before behavior stabilizes. Moreover, it sometimes becomes clear in the course of an investigation that stability criteria that appeared reasonable when a study was designed cannot, in fact, be met, or that they allow for so much variability in behavior that possible effects of the independent variable are obscured. In either case, it is appropriate to change the criteria. Care must be taken, however, to ensure that the change is (1) actually warranted and (2) duly noted in descriptions of the study.

A second consideration in the use of withdrawal designs is to change only one variable at a time. This convention actually holds for all designs and characteristically is violated only when two or more treatments are being compared. Consider a situation in which a researcher is interested in the combined effects of two treatments, B and C. Their effects could be assessed appropriately via an A–B–A–B–BC–B–BC design (BC indicates that both treatments are in effect simultaneously). Although this configuration does not provide for an analysis of the C treatment alone, it does allow the combined effects of the two treatments to be examined by comparing performance between the adjacent B and BC phases, which differ with respect to a single variable (C) only.

Withdrawal designs have three limitations, two of them major. One is that, for obvious reasons, such designs cannot be used to evaluate independent variables that produce irreversible effects.

A second limitation in applied settings is that withdrawing an apparently effective treatment, and thereby inducing countertherapeutic behavior change, may be practically or ethically unacceptable. If, for instance, the A–B phases of a proposed A–B–A design were to suggest that an intervention substantially reduced pernicious self-injury by a child, it would be difficult to justify returning to baseline just to provide further support for this possibility. Similar reasoning would apply if an incentive plan appeared to increase sales in a struggling business.

A third potential limitation of withdrawal designs is that, characteristically, many sessions are required to complete an experiment. This limitation may be a problem in some applied settings, in which subjects are available for a limited time only.

Multiple-Baseline Designs

In the multiple-baseline design, a number of target behaviors, typically three or four (Kazdin & Kopel, 1975), are recorded. These behaviors can be different behaviors of a single subject, the same or different behaviors of two or more subjects, or the same behavior of a single subject in different situations. Each target behavior must require change in the same direction, and all dependent measures should be independent of each other (i.e., changing one ought not to affect the others).

The multiple-baseline design typically begins with an assessment of all target behaviors during baseline conditions.* Once performance is stable, treatment is introduced for one target behavior. For example, if the aggressive responses of three different children constitute the target behaviors, one child's aggressive responding would be treated, whereas the aggressive behavior of the other two children would continue to be recorded under baseline conditions. When the behavior first treated had stabilized in the presence of the intervention, treatment would begin for a second behavior, and continue for the first. Treatment for the third target behavior would begin only when the second had stabilized. Data showing

---

*The multiple-baseline design can also begin with treatment of all target behaviors, after which treatment is withdrawn in a sequence that is temporally staggered across target behaviors. This arrangement is used, of course, to evaluate the effects of withdrawing a treatment.

the results of hypothetical experiment in which a multiple-baseline-across-subjects design was employed are presented in Figure 5.1.

With a multiple-baseline design, a treatment's effectiveness is evident if each dependent measure changes when and only when treatment is initiated for that behavior. If two or more behaviors are apparently affected when treatment is implemented for one of them, this outcome cannot be confidently attributed to the treatment. Although such an outcome might reflect a nonindependence of the target behaviors, such that changing one also changed the others, it might also reflect the action of some extraneous variable that affected all the target behaviors. These alternative possibilities sometimes can be teased apart by terminating treatment and determining whether all target behaviors return to pretreatment levels. If they do, it is reasonable to conclude that treatment was effective and the two behaviors are nonindependent. If not, the action of an extraneous variable cannot be ruled out, and the efficacy of the intervention remains moot.

Poche, Brouwer, and Swearingen (1981) provide an example of the use of a multiple-baseline-across-subjects design in a study intended to evaluate a procedure for developing self-protective behavior in young children. Results of the investigation are shown in Figure 5.3.

Prior to training, which involved modeling, behavior rehearsal, and social reinforcement, all three subjects were susceptible to adult lures in that they remained in close proximity to, and agreed to go with, the adult. Following training, each subject responded appropriately to lures (e.g., moved away and said "No" when asked to accompany the adult). Follow-up data collected 4 months after treatment had ended indicated that treatment gains were well maintained over time.

The multiple-baseline design is not limited by the two major shortcomings of withdrawal designs. It is appropriate for evaluating treatments that produce irreversible effects (so long as the dependent measures are independent of each other) and does not require countertherapeutic behavior change to demonstrate the efficacy of treatment. For these reasons, the design is quite popular in applied behavior analysis. It is rarely used, however, in basic research.

There are four potentially significant limitations of the multiple-baseline design. One is that withholding treatment during an extended baseline period may be ethically or practically undesirable. A second is that the design is appropriate only when the behaviors of concern are independent, and as Kazdin and Kopel (1975) note, independence of target behaviors is sometimes difficult to determine prior to the initiation of treatment. A third is that it may be difficult to interpret the efficacy of an intervention when it produces the desired change in some target behav-

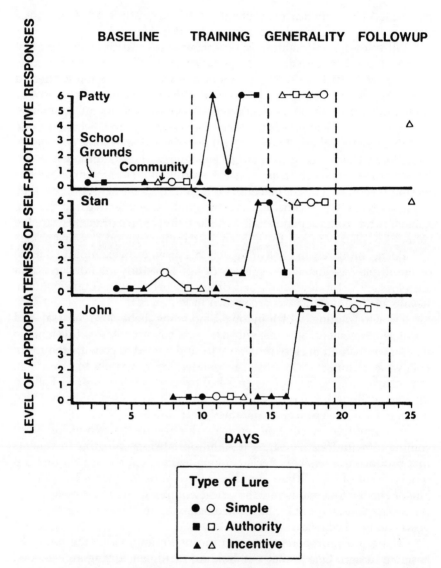

FIGURE 5.3. Level of appropriateness of self-protectiveness during baseline, training, and generality probes in both school and community settings. (●, ■, ▲) Data gathered near the school (where children were trained); (○, □, △) data gathered in a location away from the school. A multiple-baseline design (with follow-up) was used in this study; the conditions are described in the text. From C. Poche, R. Brouwer, and M. Swearingen, "Teaching self-protection to young children," *Journal of Applied Behavior Analysis*, Vol. 14, pp. 169–176. Copyright 1981 by the Society for the Experimental Analysis of Behavior. Reprinted by permission.

iors, but not in others. A fourth is that the design, unless elaborated, is not well suited for evaluating treatment interactions, comparing interventions, or examining multiple values of an independent variable.

Multiple-Probe Designs

The multiple-probe design (Horner & Baer, 1978) combines the logic of the multiple-baseline design with probe procedures to evaluate the effects of interventions designed to teach a sequence (chain) of responses. With this design, initial baseline data are collected on each of the steps (behaviors) in the training sequence. After initial baseline data are collected, treatment is introduced for the first behavior in the chain. Only when that response is acquired to a criterion level are additional probe data collected for other behaviors in the sequence. These data represent a "true baseline" for the second response in the sequence, which is now exposed to treatment. At the end of treatment for this second behavior, probe data are collected for all other responses; these data constitute follow-up data for the behavior initially treated and true baseline data for the third behavior in the sequence. The logic of the multiple-probe design and its relation to the multiple-baseline are evident in Figure 5.4, which shows data collected by Braam and Poling (1983) in a study examining the use of transfer of stimulus control procedures to teach intraverbal responses to mentally retarded participants.

Prior to training (during the baseline phase), the subject did not emit appropriate intraverbal responses (i.e., spoken words representing specific examples) in response to the generic nouns "color," "school," "vehicle," "drink," and "people." Appropriate intraverbal responding to each of these nouns developed rapidly with exposure to a transfer of stimulus control procedure (during the training phase), and probe data collected after training ended indicated that treatment gains were maintained over time.

As does the unelaborated multiple-baseline design, the multiple-probe design demonstrates the efficacy of treatment by showing that a particular behavior changes when and only when treatment for that behavior is instituted. The general strengths and weaknesses of the two designs are precisely the same. Because the multiple-probe procedure does not require baseline observations throughout the course of a study, it is of real value when "continuous measurement during extended multiple baselines proves impractical, unnecessary, or reactive" (Horner & Baer, 1978, p. 196).

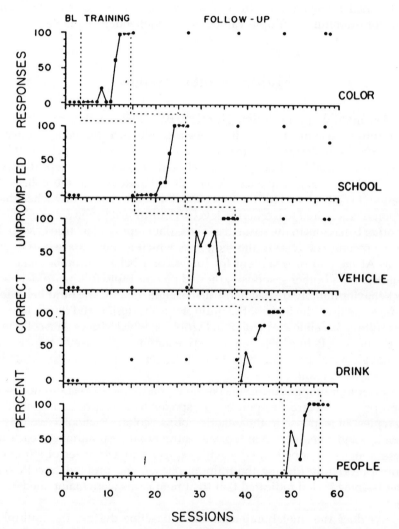

Figure 5.4. Percentage of correct unprompted intraverbal responses by a mentally retarded subject. A multiple-probe design was used in this study. Training began with immediate prompts (▲), followed sequentially by sessions in which prompts were delayed by 1 sec (●) and by 2 sec (■). (BL) Baseline. From S. Braam and A. Poling, "Development of intraverbal behavior in mentally retarded individuals through transfer of stimulus control procedures: Classification of verbal responses," *Applied Research in Mental Retardation*, Vol. 4, pp. 279–302. Copyright 1983 by the Society for the Experimental Analysis of Behavior. Reprinted by permission.

Other Probe Designs

In some experiments, brief exposures to an independent variable, termed *probe manipulations*, are superimposed on otherwise constant conditions. As Perone (1991, p. 157) indicates, "Perhaps the most common use of such designs today is to study the effects of drugs on schedule-maintained responding." A study by Picker and Poling (1984), introduced in Chapter 2, illustrates the use of a probe design in this general context. In this experiment, the independent variables were five anticonvulsant drugs (phenobarbital, clonazepam, valproic acid, ethosuximide, and phenytoin), and each was evaluated across a range of doses (values) in pigeons responding under a repeated acquisitions procedure. Under this assay, which measures learning, the birds were required to peck three response keys in a spatially defined sequence that varied each day. The percentage of responses that were errors was one dependent variable; response rate was the other.

With protracted training in the absence of drug, the number of errors made by a subject in mastering each new sequence became relatively stable and provided a sensitive baseline against which drug effects could be assessed. This assessment was made by giving each drug and dose of interest on two occasions, with the order of administration determined at random for each subject. Drug sessions were separated by at least three control sessions in which responding was stable and vehicle injections were given.

As shown in Figure 5.5, which presents data for a single bird, all the drugs decreased response rates at certain doses, indicating that the doses used were behaviorally active. With the exception of ethosuximide, each drug also increased errors (impaired learning) relative to control values. These results suggest that there are qualitative as well as quantitative differences in the effects of anticonvulsant drugs on learning, a finding that may have implications for their use in treating epilepsy in humans. The results also are consistent with preclinical and clinical data indicating that anticonvulsants can adversely affect learning, although this effect is not a major problem with ethosuximide at therapeutic doses.

Probe designs are adequate for demonstrating functional relations between an independent and a dependent variable, and they yield much information in relatively little time. Such designs are inappropriate for evaluating interventions with irreversible or long-lasting effects or interventions that produce behavioral effects only with protracted exposure. For example, antidepressant drugs produce clinical benefits only after 2 or 3 weeks' exposure (Poling et al., 1991) and could not be fairly evaluated with a probe design.

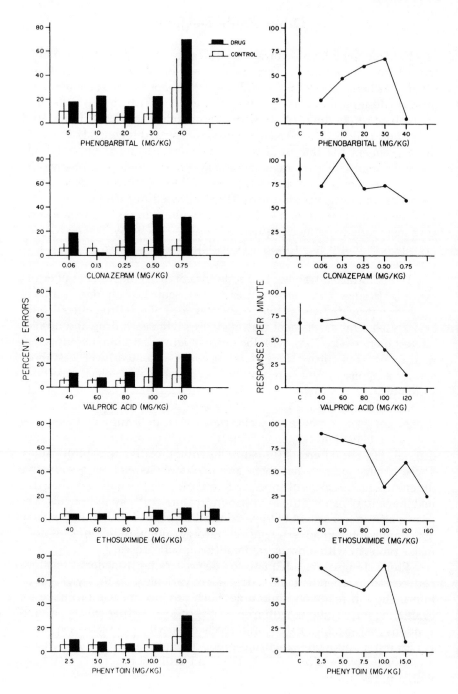

Changing-Criterion Designs

Although sometimes described as a variant of the multiple-baseline design (e.g., Hartmann & Hall, 1976), the changing-criterion design is perhaps better envisioned as an A–B design with the treatment phase divided into subphases, each of which involves a different criterion for reinforcement. As Hartmann and Hall (1976, p. 527) noted:

> Each treatment phase is associated with a stepwise change in criterion rate for the target behavior. Thus, each phase of the design provides a baseline for the following phase. When the rate of the target behavior changes with each stepwise change in the criterion, therapeutic change is replicated and experimental control is demonstrated.

Figure 5.6 shows for one subject the results of a study by Foxx and Rubinoff (1979) in which a changing-criterion design was used to evaluate the effectiveness of a procedure designed to reduce the caffeine intake of heavy coffee drinkers. Prior to treatment, which involved self-monitoring and contingency contracting, this subject consumed over 1000 mg caffeine per day. Each of four successive treatment phases reduced the amount of caffeine allowed by 100 mg relative to the previous phase. Results indicate that the subject met the criterion at each phase of treatment and that the reduced caffeine intake associated with treatment was retained over a prolonged follow-up period.

The changing-criterion design has three major strengths: First, treatment does not have to be withdrawn; thus, ethical and practical problems related to countertherapeutic behavior change are obviated. Second, all subjects receive treatment after only a brief baseline period. Third, when performance closely matches specified criteria, the design allows for an unambiguous demonstration of a treatment's efficacy.

Problems of interpretation arise, however, when behavior does not closely parallel criterion levels, which is one of the design's shortcomings. A second shortcoming is that the design does not allow for the evaluation of treatments that do not specify stepwise changes in performance. Such treatments include most interventions of interest to behavior analysts.

←——

FIGURE 5.5. Effects of phenobarbital, clonazepam, valproic acid, ethosuximide, and phenytoin on response rate and percentage errors for one pigeon exposed to a repeated acquisition procedure. *Left*: (□) Mean percentage errors during control sessions; the vertical lines represent the ranges across these sessions. Percentage errors for control sessions reflect performance during predrug sessions until a number of reinforcers equivalent to that obtained during the following drug session was obtained. (■) Mean percentage errors during initial exposure to the listed drug and dose. *Right*: (C) Mean rate of responding during control sessions; the vertical lines represent the ranges across these sessions. Redrawn from Picker and Poling (1984).

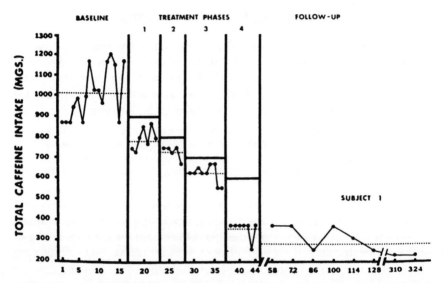

FIGURE 5.6.   One subject's daily caffeine intake during baseline, treatment, and follow-up phases. A changing-criterion design was used in this study, and treatment involved self-monitoring and contingency contracting. From R. M. Foxx and A. Rubinoff, "Behavioral treatment of caffeinism: Reducing excessive coffee drinking," *Journal of Applied Behavior Analysis*, Vol. 12, pp. 335–344. Copyright 1979 by the Society for the Experimental Analysis of Behavior. Reprinted by permission.

Alternating-Treatments Designs

The alternating-treatments design (Barlow & Hayes, 1979) involves repeated measurement of behavior while conditions rapidly alternate, typically between baseline and a single intervention phase, or between two separate intervention phases. Conditions may alternate either within a measurement session or from one session to the next, and the sequencing of conditions may be either regular or random. In many cases, a unique exteroceptive stimulus is paired with each condition. Such an arrangement, which Ulman and Sulzer-Azaroff (1975) term a "multielement baseline design," somewhat resembles a multiple-baseline design (see Figure 5.1).

The logic of the alternating-treatments design is evident in Figure 5.7, which presents data collected by Mosk and Bucher (1984). They demonstrated that shaping plus prompting was more effective than prompting alone in teaching visual–motor (pegboard) skills to moderately and severely mentally retarded children.

One alleged advantage of the alternating-treatments design is in the analysis of highly variable behaviors. In applied settings, levels of a target

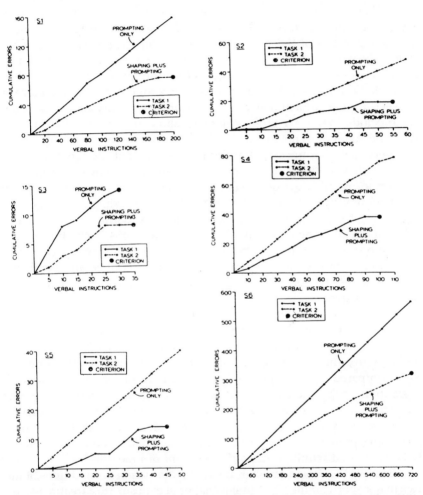

FIGURE 5.7. Cumulative errors in a pegboard-insertion task for subjects S1–S6. Cumulative errors are graphed as a function of the number of instructions; the larger circles show where criterion was reached. An alternating-treatments design was used in this study to compare a prompting procedure to a procedure in which prompting was combined with shaping. From M. D. Mosk and B. Bucher, "Prompting and stimulus shaping procedures for teaching visual-motor skills to retarded children," *Journal of Applied Behavior Analysis*, Vol. 17, pp. 23–34. Copyright 1984 by the Society for the Experimental Analysis of Behavior. Reprinted by permission.

behavior often fluctuate widely across time; behavior may also consistently improve or worsen in the absence of treatment. In such cases, a phase change is not appropriate when multiple-baseline or withdrawal designs are employed. With the alternating-treatments design, conditions change regardless of the subject's behavior, and a comparison can legitimately be made between performance in two conditions (e.g., treatment and baseline) even though the target behavior improves or worsens during each. So long as behavior is consistently and appreciably better (or worse) during treatment than during baseline, variability across time does not preclude making a gross statement about the effects of treatment. When behavior does not inevitably differ across conditions, the appropriate interpretation is unclear; the alternating-treatments design deals effectively with variability only when interventions produce effects large enough to overshadow other sources of variability. The effects of such interventions typically are evident regardless of experimental design. The alternating-treatments design differs from other within-subject designs primarily in allowing phase changes while behavior is fluctuating; it cannot impose order on chaotic behavior. Moreover, one anticipates that a truly useful applied intervention will not only produce an overall improvement in behavior, but also be associated with relatively stable performance over time.

A second advantage of the alternating-treatments design is its relative efficiency. Unlike withdrawal or multiple-baseline designs, the alternating-treatments design allows for early initiation of treatment (although an initial pretreatment baseline phase is common when two interventions are to be compared), rapid exposure to all conditions of concern, and a quick evaluation of the success of treatment. Meaningful data (i.e., data that allow for a comparison of behavior under all the conditions of interest) are generated early in the experiment, so that all is not lost if the project terminates prematurely. In addition, the design allows diverse treatments to be compared within a reasonable amount of time.

Because the design involves the rapid alternation of conditions, it has two major shortcomings. The first is that brief exposure to a treatment may be insufficient for its true actions to be observed. The second is that the design is inappropriate for evaluating treatments with long-lasting effects, because their actions will persist into, and confound, subsequent conditions.

Concurrent-Schedule Designs

Under a concurrent schedule, two or more alternative schedules of reinforcement are simultaneously and independently in effect for two or more different responses (Catania, 1992). By determining the amount of time or responding allocated to one alternative relative to the other, the variables that influence choice, or preference, can be isolated. A substantial literature exists in this area; much of it attempts to derive mathematical formulae that precisely relate environmental inputs to environmental outputs (e.g., Davison & McCarthy, 1988; de Villiers, 1977).

A study by Herrnstein (1961), introduced in Chapter 2, illustrates the use of concurrent schedules to study choice. He exposed three pigeons to conditions in which concurrent variable-interval (VI) schedules of food delivery were arranged on two response keys. In most conditions, a 1.5-sec changeover delay (COD) was arranged. The COD ensured that no food could be delivered until at least 1.5 sec had elapsed from the time a bird switched from pecking one key to pecking the other. Several different VI combinations were compared. Each VI schedule specified that food became available aperiodically, with the average time between successive food availabilities equal to the schedule value (e.g., 60 sec under a VI 60-sec schedule), and was delivered dependent on a response. Herrnstein recorded the number of pecks emitted and the number of food deliveries obtained under each alternative. When these measures were compared across schedule combinations, the relative proportion of responses emitted under a schedule matched the relative proportion of reinforcers (food deliveries) under that alternative. These findings are depicted in Figure 5.8.

Variables with relatively small effects in other situations, including duration of reinforcement and delay of reinforcement, often produce much larger effects when concurrent schedules are arranged. Therefore, such schedules are quite useful for studying such variables, as well as others that influence choice (Catania, 1992).

Extraneous Variables and Experimental Design

As discussed in Chapter 2, the most important consideration in selecting a design is to ensure that the arrangement of conditions is adequate to permit reasonable inferences about any effects that an independent variable may have on a dependent variable. Reasonable inferences can be made only when conditions are arranged so that the effects of the independent variable are not likely to be confounded with those of known or anticipated extraneous variables. In many cases, a significant aspect of experimental design is arranging conditions so as to separate the effects of a potential, and specifiable, extraneous variable from those of the independent variable. Separating these effects may require the use of experimental

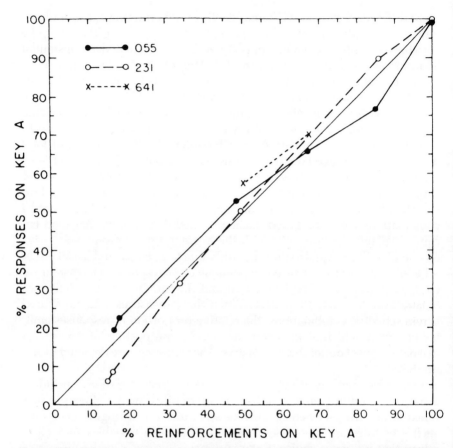

FIGURE 5.8. Relative frequency of responding to one alternative under concurrent VI VI schedule as a function of the relative frequency of reinforcement for that alternative. The diagonal line indicates "matching," where relative response rates approximate relative reinforcement frequencies. From R. J. Herrnstein, "Relative and absolute strength of response as a function of frequency of reinforcement," *Journal of the Experimental Analysis of Behavior,* Vol. 4, pp. 267–271. Copyright 1961 by the Society for the Experimental Analysis of Behavior. Reprinted by permission.

arrangements not readily described by the conventional nomenclature of experimental design.

A study by Poling and Thompson (1977) provides an example of the use of a specific experimental arrangement, yoked-control subjects, to tease apart the effects of an independent variable and those of a potential extraneous variable. This study examined whether ethanol-reinforced responding by rats could be suppressed by delaying the availability of food. Initially, food was available under a fixed-interval 26-sec (FI 26-sec) schedule* and 8% ethanol solution was available under a fixed-ratio (FR 1) schedule. This condition was in effect until the rate of ethanol-reinforced responding for each of three rats showed no obvious trend across 7 consecutive 1-hr sessions. When this criterion was met, conditions were changed so that each ethanol-reinforced response postponed food availability by 8 sec. Imposing the 8-sec delay substantially reduced ethanol-reinforced responding in each of the three rats. In subsequent conditions, similar results were observed with ethanol concentrations of 0%, 8%, 16%, and 32%. In the last part of the study, when 8% ethanol was available, delays of 8–2048 sec suppressed ethanol-reinforced lever pressing and ethanol consumption relative to baseline (no delay) levels, although degree of suppression did not increase monotonically with delay length. These results suggest that delaying food availability dependent on ethanol-maintained responding punished such responding.

A tenable alternative explanation of these results is possible. It is well established that eating influences drinking in rats and that water intake is related to the interfood interval (e.g., Falk, 1967, 1971). These relations having been established, it is possible that the reduced drinking observed by Poling and Thompson (1977) resulted from changes in interfood interval per se, not from the delay dependency. To evaluate this possibility, each of the three "master" rats described above was matched with a yoked-control partner. Food became available for a yoked-control rat as soon as it was earned by the master partner and was delivered dependent on a lever press by the yoked-control rat. Ethanol solution of the same concentration as that available to the master partner was always available to the yoked-control rat under an FR 1 schedule. Ethanol-maintained lever presses by yoked-control rats never affected food availability. Thus, they were exposed to approximately the same interfood interval as their master partners, but they were not exposed to the delay dependency.

If changes in interfood interval, not the delay dependency, were primarily responsible for changes in ethanol intake, similar levels of ethanol-maintained responding should have been observed in master rats and their yoked-control partners. Such similarities were not observed. As Figure 5.9 shows, responding by yoked-control animals was not reduced under conditions in which ethanol-reinforced responding by their master partners delayed food availability. Thus, the delay dependency, not changes

---

*Why FI 26-sec? A mistake, really. The intended value was FI 30-sec, but training had begun before a programming error was detected. Did it make a difference? Probably not.

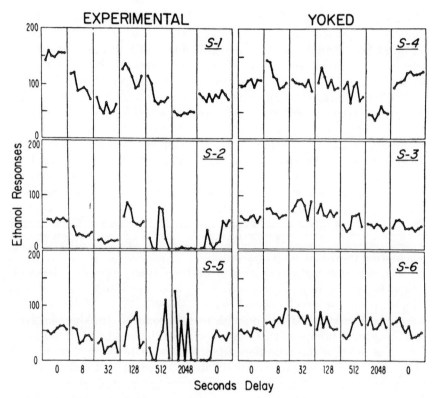

FIGURE 5.9.   Ethanol-reinforced lever presses per session when food was concurrently available and each lever press reinforced by an 8% ethanol solution delayed food availability 0–
2048 sec. Each ethanol-reinforced lever press by an experimental animal delayed food
availability by the number of seconds indicated on the abscissa. The baseline schedule of food
availability was FI 26-sec, and all sessions were 1 hr in length. From A. Poling and T.
Thompson, "Suppression of ethanol-reinforced lever pressing by delaying food availability,"
Journal of the Experimental Analysis of Behavior, Vol. 28, pp. 271–283. Copyright 1977 by the
Society for the Experimental Analysis of Behavior. Reprinted by permission.

in the interfood interval, appeared to be responsible for the response
suppression observed in the master animals.

Good experimental strategies reduce to an acceptable level the probability that the effects of extraneous variables will be confounded with those
of independent variables. But no single study attempts to examine or
control for all the variables that might influence results. Consider, for
example, the possibility that the results obtained by Poling and Thompson
(1977) were to some extent influenced by order effects. *Order effects* are

evident when the effects of particular interventions (or intervention levels) are influenced by the order in which a subject is exposed to them, and are a possibility whenever two or more experimental conditions are of interest. In the first phase of the study by Poling and Thompson (1977), the effects of delaying food availability by 8 sec dependent on liquid-maintained responding was examined, in order, when the liquid contained 8%, 0%, 8%, 16%, and 32% ethanol. The degree of response suppression produced in the rats exposed to the delay dependency was greatest at the 8% concentration and least at the 32% concentration. Is it possible that the effects of the 8-sec delay at a particular ethanol concentration depended, in part, on prior exposure to that delay at other concentrations?

It certainly is, although the authors made no attempt to evaluate this possibility systematically. Instead, in subsequent phases of the study, they demonstrated that the effects of the 8-sec delay at the 8% concentration could be replicated. This demonstration provided a degree of assurance that the effects of the delay procedure were relatively robust and were not strongly modulated by the order of exposure to ethanol concentrations. More compelling proof could have been provided by replicating results at each of the ethanol concentrations, but no attempt was made to do so. Instead, the authors went on to examine the effects of longer delays at a single (8%) ethanol concentration.

There are no absolute standards for determining the extent to which a researcher should be concerned with order effects or other potential confounds or the steps that should be taken to deal with them. *As a rule, extraordinary claims demand extraordinary proofs.* That is, the methodology employed by researchers working in new and important areas, or those who report findings contrary to the norm, will be exposed to especially close scrutiny by the scientific community. Such scrutiny may also be applied, albeit to a lesser degree, to the methodology used by scientists without a good track record. A bit of good advice for young researchers is: "Be careful." A conservative approach to research, one that arranges more replications and controls more extraneous variables than are absolutely necessary, is likely to yield results that others can reproduce. Reproducible results are ultimately the only results of value to a science of behavior—or to any science.

Summary and Conclusions

The best way to learn about the variables that control behavior is to use within-subject experimental designs with repeated and direct measures of behavior. There are many ways to arrange such experiments;

several common configurations were discussed in this chapter. With the exception of case study and A–B designs, which are nearly worthless, all the design variations can provide useful information about behavior and the variables that control it. Learning the logic, strengths, and weaknesses of general design strategies is prerequisite for designing successful studies.

There is no substitute for experience—ideally, experience under the tutelage of a successful researcher—in learning how to design successful studies. Nonetheless, careful reading of Sidman (1960) and Johnston and Pennypacker (1993a,b) can prove invaluable in gaining general advice about useful tactics and strategies of behavior analytic research. Those authors do an exceptional job of explaining the logic and advantages of within-subject experimentation, although their texts are not easy reads. Useful specific advice on within-subject research designs appropriate for applied behavior analysis is available in Kazdin (1982) and Barlow and Hersen (1984); both of these texts are easy to read. Researchers interested in the experimental analysis of behavior, especially with nonhuman subjects, will benefit from the information in two books edited by Iverson and Lattal (1991a,b). These books provide written coverage of many details of experimental procedures that historically were passed informally from mentor to student as lab lore. Aspiring researchers will sharpen their design skills by consulting all the sources mentioned above, reading journal articles in their area of interest (paying careful attention to the details), and consulting with established scientists.

CHAPTER 6

Between-Subjects Designs and Nonexperimental Methods

Between-subjects designs provide alternatives to within-subject strategies for arranging experimental conditions. As Kazdin (1982) points out, between-subjects designs are not favored by behavior analysts. Johnston and Pennypacker (1993b), for example, assert that between-subjects designs do not obtain useful information on functional relationships, nor do they result in findings that are meaningfully generalized to individuals who have not actually been studied. Criticisms of between-subjects approaches to research are based on the manner in which the independent variable is manipulated, how often dependent variables characteristically are measured, and how data usually are analyzed.

In experiments using between-subjects designs, different subjects are exposed to different levels of the independent variable. In most such studies, the dependent variable is measured relatively few times for each subject, and data are analyzed through the use of inferential statistics that compare group means. These tactics are less than optimal for providing information about the variables that control behavior or for evaluating the response of individual people to therapeutic interventions. Nonetheless, between-subjects experimental designs are occasionally useful for behavior analysts. Mixed within-subject and between-subjects designs are even more likely to be useful, and they play an important role in behavior analysis. The purpose of this chapter is to consider the logic of between-subjects experimental designs and to introduce some common between-subjects experimental configurations. *As a rule, within-subject designs are preferable to between-subjects designs, and the latter should be used only when there is good reason to do so.* For this reason, our coverage of between-subjects designs is brief. Between-subjects designs are covered at length in traditional experimental design and statistics books (e.g., Campbell & Stanley, 1963; Howell, 1992; Isaac & Michael, 1981) and characteristically are a major focus of college courses dealing with these topics.

Reasons for Using Between-Subjects Designs

The best reason for using between-subjects designs is that the experimental question cannot be answered through the use of a within-subject design. In such cases, a between-subjects design is used by default. For example, Tudor and Bostow (1991, p. 362) supported their use of a between-subjects design in a study of computer-programmed instruction by indicating that within-subject designs were inappropriate for their purposes:

> Cumulative instructional effects ruled out a reversal design. A multiple base-line design could not be used because students could easily converse with each other and make comparisons, the effects of which could not be controlled or evaluated. Also, the content of the instructional program could not be broken into truly independent and equal segments enabling sequential training comparisons.

Thus, Tudor and Bostow (1991) opted for a between-subjects design because of the lack of a viable alternate within-subject design.

A second reason for using between-subjects designs is that one wishes to compare interventions that have been previously verified as effective through within-subject experiments (Ballard, 1986; Barlow & Hersen, 1984). Although some interventions can tenably be compared on a within-subject basis, those that produce irreversible changes in behavior cannot. For example, Wilkenfield, Nickel, Blakely, and Poling (1992) were interested in response acquisition with delayed reinforcement. There is no generally accepted way to arrange delayed reinforcement; several procedures are available, and they do not have identical effects on behavior. Recognizing this difficulty, Wilkenfield et al. (1992) compared the rapidity with which experimentally naive rats acquired a lever-pressing response under three different procedures involving delayed reinforcement. Under one procedure, every response produced food after a specified delay. Under a second procedure, the first response produced food after a specified delay, and responses during the delay reset that delay. Under a third procedure, the first response produced food after a specified delay, and responses during the delay had no scheduled consequences. During 8-hr sessions, delays of 4, 8, and 16 sec were arranged under each of the three procedures. Nine different rats were exposed to each condition of interest.

The performance of rats that never received food and of rats that received food immediately following each response was also evaluated. The primary dependent variable was the cumulative number of responses emitted across time.

Results indicated that responding was acquired within 2 hr at all delays under all procedures, although the cumulative response functions differed as a function of the delay value and the procedure used to arrange the delay. Repeated measures of the behavior of individual subjects were taken in this study, and data analysis was primarily graphic, but the design was a between-subjects configuration. Figure 6.1 presents the data obtained at various delay values under one of the three delay procedures.

The study by Wilkenfield and colleagues illustrates an important point: *Even when a between-subjects design is used, repeated measures and graphic analysis of the behavior of each subject may be possible, and valuable as well.* A problem with many studies using between-subjects designs is that much potentially important information concerning the behavior of individual subjects is lost, but this problem can be minimized with careful forethought.

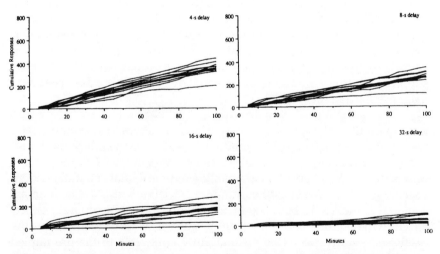

FIGURE 6.1. Cumulative responses across the first 100 min of experimental sessions for experimentally naive rats exposed to procedures in which a lever press produced food following the indicated delay and responses during the delay reset the delay interval. Nine rats were exposed to each delay. (— — —) Data for an individual rat; (———) group mean. From J. Wilkenfield, M. Nickel, E. Blakely, & A. Poling, "Acquisition of lever-press responding with delayed reinforcement: A comparison of three procedures," *Journal of the Experimental Analysis of Behavior*, Vol. 58, pp. 431–443. Copyright 1992 by the Society for the Experimental Analysis of Behavior. Reproduced by permission.

A third reason for using between-subjects designs is that the audience demands such designs. For instance, some funding agencies appear to be especially likely to support research examining clinical procedures that have been shown to be effective in conventional between-subjects designs using statistical data analysis (Ballard, 1986; Barlow & Hersen, 1984). A researcher desirous of funding from such agencies would be wise to conduct some between-subjects evaluations.

There may be other reasons, of course, for using between-subjects designs. Before beginning any experiment, one must determine what evidence is required to demonstrate the presence or absence of a treatment effect and how conditions must be arranged to collect such evidence. Occasionally, the behavior analyst will conclude that a between-subjects design is the best option for achieving these ends.

The Logic of Between-Subjects Designs

The logic of all between-subjects experimental designs is similar and can be illustrated by considering the simplest between-subjects design, a *randomized two-groups design*. With this design, two levels of an independent variable are of interest. Subjects in one group are exposed to one level of the independent variable and subjects in a second group are exposed to the other level. Such a design might be appropriate, for example, for addressing this experimental question: Does treatment with the appetite-suppressant drug fenfluramine decrease body weight in adolescent girls diagnosed as obese?

An important initial step in addressing this question is selecting subjects. In traditional approaches to experimental design, strong emphasis is placed on selecting subjects at random from the larger group of interest, termed a *population*, to create a representative sample. *Random selection* means that all members of the population have an equal and independent chance of being selected. When random selection is used, it is often assumed that the results obtained in a study may be unambiguously generalized to the population from which the subjects were drawn. That is, the results of a study would be meaningfully applicable to the unsampled members of the larger group (population) had they also been exposed to the treatment manipulation. Figure 6.2 shows the random selection procedure.

Assume that in our example, the population of interest comprises all the females between 12 and 18 years of age being treated for obesity at a particular weight-loss clinic. There are 137 people in this clearly defined population, too many to be studied individually. The clinic director there-

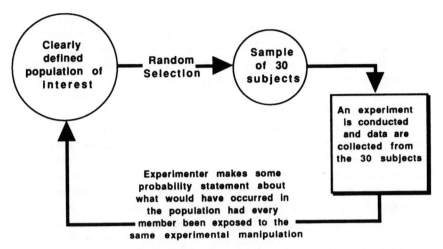

FIGURE 6.2. Schematic depiction of the random selection procedure. In the example from the text, the population comprises 137 obese adolescent females and the sample comprises a subset of 30 of them. Each member of the population had an equal and independent chance of being selected as part of the sample.

fore decides to study a subset (sample) of this population to evaluate the effectiveness of the pharmacological intervention. The researcher draws at random from the names of all the adolescents to isolate a sample of, for example, 30 subjects.

Subjects from this initial sample are then randomly assigned to either of two groups, each comprising 15 people. Members of the experimental group will receive a 20 mg tablet of fenfluramine (Pondimin) before each meal (60 mg per day); members of the control group will receive a placebo tablet before each meal. In all other respects, the two groups will be treated as similarly as is possible. In *random assignment*, all subjects have an equal and independent chance of being assigned to a particular group. Random assignment produces groups that should be similar at the onset of a study. This procedure generally ensures that extraneous variables are evenly distributed across the groups. Thus, it is unlikely that the various groups will differ significantly with respect to levels of the dependent variable. In the case of our hypothetical study, for example, body weights for subjects in the two groups should be similar prior to the intervention.

At some point in our hypothetical study, perhaps after 2 weeks, body weights of all subjects will be measured and compared through statistical analysis of group means. The analysis begins with the assumption that the means of the populations from which treatment and control groups were selected are equivalent. Because the treatment and control groups are merely subsets of the population, however, there will almost always be differences. Statistical tests (described further in Chapter 8) tell us the probability of finding sample mean differences of at least the magnitude obtained if there had been no differences between the means for populations of treated and untreated subjects. If the probability is sufficiently low, usually less than 0.05, it is concluded that the intervention was effective.

In our hypothetical study, assume that at the end of the experimental period, the mean weight of subjects in the treatment group was 193 pounds and that of subjects in the control group was 207 pounds. Assume, moreover, that these group means were compared via an appropriate inferential statistic (e.g., a t-test) and that the difference in means is significant. Given these findings, it may be concluded that the drug treatment was effective.

What can be said concerning the 107 adolescents not selected for the study? Because random selection was employed, we could say that the results are generalizable to the larger group. That is, we can infer that similar results would obtain for unsampled persons in the larger group (i.e., the original 137 adolescent girls from which the sample was drawn) if they were exposed to the same treatment manipulation.

Random sampling is not always possible, but other sampling procedures may be used. One such procedure is *convenience sampling*, which employs an available pool of people as participants in a study. For example, 75 undergraduate students from a psychology course served as subjects in a study conducted by Tudor and Bostow (1991). Students received a grade of "pass" for participating in the study as a part of their course requirement. The students enrolled in the psychology class constituted a convenience sample of available participants. With convenience sampling, all individual members of a population (e.g., all students enrolled at the university at which the research is conducted) do not have an equal and independent chance of being selected, and the population to which the results are said to generalize is ambiguous. To whom, then, can the results be extended in the Tudor and Bostow example? They can be extended to a group of students having characteristics similar to those of the students enrolled in the psychology class from which the subjects were drawn. But because the class was not explicitly defined and sampled from, those characteristics cannot be clearly specified.

The logic of between-subjects experimental designs is straightforward and compelling in principle. Problems abound, however, in applying that logic. Six are noteworthy:

1. *It is often difficult to identify populations of interest or to gain equivalent access to all members of those populations.* A researcher usually works with

readily available subjects, although she or he is interested in extending findings well beyond those subjects.

2. *Even when populations can be defined and samples taken from them, it is never certain that results based on samples actually generalize to populations.* As discussed in Chapter 2, the only way to ascertain whether results of any one study are valid is to replicate them. This caution holds regardless of the experimental design used, but is sometimes overlooked by researchers who place undue faith in the logic of extrapolating results to populations from samples.

3. *Practical or ethical considerations often constrain or prohibit selecting subjects randomly or assigning them randomly to treatment or control groups.* Although there are strategies for dealing with such limitations, their use compromises the elegance and simplicity of the basic between-subjects arrangement.

4. *Studies that emphasize mean group differences are likely to ignore important aspects of the data.* This problem relates, of course, to data analysis strategies, not experimental designs per se, but is especially prevalent in between-subjects experiments.

5. *The manipulations made in studies that depend on after-the-fact statistical tests to evaluate findings are usually rigidly fixed and not determined on the basis of incoming data.* As discussed in Chapters 5 and 7, studies that involve within-subject designs and graphic data analysis are likely to be flexible and data-driven, which is advantageous in delineating the variables that control the behavior of individual organisms.

6. *It may be impracticable to secure the large number of subjects needed to conduct between-subjects experiments.* In some cases, sufficient numbers of subjects to use between-subjects designs are simply unavailable.

The significance of these problems varies, depending on the experimental question being addressed and the specific design being used to address it. Many different between-subjects designs have been described (e.g., Campbell & Stanley, 1963; Isaac & Michael, 1981). The terms used to describe specific between-subjects designs vary, depending on which textbooks one reads. It is convention, however, to differentiate two broad design categories, single-factor and multiple-factor. A *factor* is an independent variable; therefore, single-factor designs examine only one independent variable, whereas factorial designs examine two or more independent variables. Factorial designs allow for treatment interactions to be evaluated. An *interaction* is "an effect attributable to the combination of variables above and beyond that which can be predicted from the variables singly" (Winer, 1971, p. 309). When an interaction occurs, the effects of one factor differ, depending on the level of a second factor.

Single-Factor Designs

All *single-factor* designs are elaborations of the randomized two-groups design, which was used to illustrate the logic of between-subjects experiments. One elaboration involves taking repeated measures of the independent variable. For example, in our hypothetical study of the effects of fenfluramine on the body weights of adolescents, it would be reasonable to weigh subjects once per week over a period of several weeks and to compare weights in the experimental and control groups across time. As emphasized previously, behavior analysts generally favor experimental arrangements that include repeated measures of behavior.

A second elaboration of single-factor designs is to include more than two levels of the independent variable. In our hypothetical study, only a single dose of fenfluramine (60 mg per day) was evaluated. It is a fundamental tenet of pharmacology that drug effects are dose-dependent; therefore, it would be reasonable to evaluate the effects of other fenfluramine doses. This evaluation could be made by adding more groups to the experiment.

A third elaboration of single-factor designs is to add a pretest to ensure that dependent variable scores for all groups of interest actually are roughly equivalent prior to introduction of the independent variable. For example, it would be perfectly reasonable in our hypothetical study to weigh all participants just prior to exposing them to treatment. If this strategy was employed and initial differences in groups were detected, these differences could be considered in the ensuing analysis and interpretation of findings. In general, it is advisable to include a pretest measure in any experimental investigation. Pretests are obviously inappropriate, however, when initial measurement of the dependent variable will influence subsequent performance.

Tudor and Bostow (1991, p. 362) used a pretest–posttest design with five groups (two control and three experimental conditions) to "... evaluate the importance of active student responding while using the microcomputer to deliver the contingencies of programmed instruction." The authors randomly assigned subjects to groups that were exposed to five types of computer-programmed instruction. In the two control conditions, instructional material was presented in a manner similar to reading from a book. Subjects in the experimental groups were required to respond actively to instructional material by thinking or typing answers. Figure 6.3 shows the mean pretest and posttest scores for subjects in the five groups. These results indicate that requiring students to respond actively to instructional materials facilitated mastery.

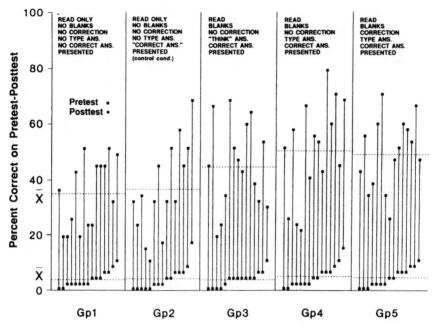

FIGURE 6.3. Percentage of correct responses made on a fill-in-the-blank pretest and posttest by groups of subjects exposed to five different types of computer-programmed instruction. Each vertical line represents the performance of a single subject; the lower data point is the pretest score and the higher data point is the posttest score. (······) Group means. From R. M. Tudor & D. E. Bostow, "Computer-programmed instruction: The relation of required interaction to practical application," *Journal of Applied Behavior Analysis*, 1991, Vol. 24, pp. 361–368.

Designs with Two or More Factors

Designs with two or more factors allow the researcher to evaluate the effects of multiple independent variables alone and together. In a two-factor design, each subject is randomly assigned to a group, and each group is exposed to a particular level of factor A in combination with a particular level of factor B. Following exposure to the treatment, measures of the target behavior are collected and analyzed. With more than two factors, groups are defined by the level of each of the factors.

An example of a design with more than one factor is provided by Poling, Kesselring, Sewell, and Cleary (1983), who examined the lethal effects of pentazocine and tripelennamine in mice housed alone and in groups. A combination of these drugs, known on the street as "Ts and blues," was popular in the early 1980s as a substitute for heroin, and several human deaths were attributed to its use. Therefore, Poling and coauthors examined the lethality of the drugs alone and in combination. Two housing conditions were examined because group housing, and the stimulation it provides, is known to increase the lethality of some drugs.

Poling et al. (1983) used a three-factor design; the factors were pentazocine dose, tripelennamine dose, and housing conditions. The lethality of 20 mg/kg tripelennamine in combination with 40, 60, and 80 mg/kg pentazocine, and that of 40 mg/kg tripelennamine in combination with 10, 20, and 40 mg/kg pentazocine, was examined under conditions of individual and group housing. The lethality of 40 mg/kg tripelennamine and 80 mg/kg pentazocine, each given alone, was also assessed under these conditions. Two randomly selected groups, each comprising 16 mice, were exposed to each of the eight drug regimens of interest (six combination doses plus a single dose of pentazocine alone and a single dose of tripelennamine alone). Under each drug condition, members of one group were housed aggregately (16 mice in one cage) after injection, and members of the other group were housed individually.

Results indicated that the lethality of pentazocine and tripelennamine combinations in mice was (1) dose-dependent, (2) potentiated relative to either drug alone, and (3) greater in group-housed than in individually housed animals. The practical significance of these findings was in providing support for the notion that Ts and blues are, in fact, a dangerous combination and in suggesting that people overdosing on the combination should not be exposed to unnecessary stimulation.

Factorial designs have two main strengths (e.g., Whalen & Henker, 1986):

1. They are effective, in that they allow one to evaluate two or more independent variables in one experiment. Doing so usually is less costly, uses fewer subjects, and is faster than conducting separate studies to evaluate each factor separately.

2. They are capable of detecting interactions between factors. The effects of one independent variable often differ depending on the level of other independent variables, and such interactions can be assessed only by evaluating the variables separately and in combination.

As Whalen and Henker (1986, p. 148) note, "If two factors interact, studies assessing only one of them provide incomplete and potentially misleading information." Consider what might happen if our hypothetical study of the effects of fenfluramine were elaborated by adding a second factor, exposure to a token economy. Assuming that two levels of the drug factor (0 mg per day and 60 mg per day) and two levels of the token economy factor (present and absent) were considered and weights were recorded on a single occasion only, the design would be a 2 × 2 factorial. Figure 6.4 shows five potential outcomes of this hypothetical study. In three cases, the treatments do not interact; that is, the effects of one treatment across levels do not differ as a function of the level of the other treatment. In two cases, an interaction is evident. These hypothetical data demonstrate that the interpretation of results from factorial designs is straightforward given that there is no interaction between the independent variables, but more problematic when an interaction occurs.

A study by Rosenfarb and Hayes (1984) demonstrates a treatment interaction in an actual clinical study. In this study, two common clinical interventions, self-statements and modeling, were compared in the treatment of children's fear of the dark. These treatments were implemented for some children in a public context, for some in a private context, and for others in a control context. Under the control and private treatment conditions, the self-statement treatment was superior to modeling in increasing the amount of time children spent in the dark. Under the public treatment condition, however, time spent in the dark was much greater for children exposed to either self-statement or modeling treatments, and the two were equally effective.

Because complex factorial designs appear to be capable of providing a great deal of information in a single study, beginning researchers are often tempted to use such designs. This temptation should be resisted. As Whalen and Henker (1986, p. 151) caution, "Although there is no theoretical limit to the number of factors that can be included in a factorial arrangement, there are numerous practical and logical limits. Usually, when one goes beyond three factors, these designs rapidly lose efficiency and interpretability." In particular, interactions among three or more variables often are difficult to interpret and are likely to be of dubious practical or theoretical significance. *As a rule, simple experiments are more likely than complex experiments to yield meaningful results and are recommended.*

Factorial Designs Employing Classification Variables

Factorial designs can be extended to include classification variables. A *classification variable* is a characteristic that preexists in a population. An experimenter does not manipulate a classification variable, either because it is impossible to do so (e.g., in the case of the sex of subjects) or because it is unethical to do so (e.g., one cannot ethically require human subjects to smoke cigarettes).

Figure 6.5 illustrates how, in a two-factor experiment with one classification and one treatment variable, subjects would be selected in relation

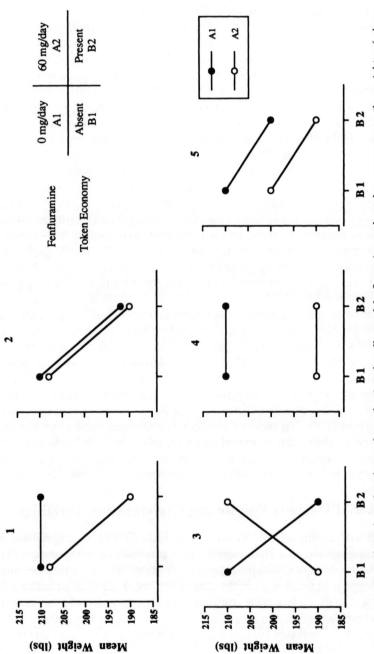

FIGURE 6.4. Hypothetical results of a study evaluating the effects of fenfluramine and a token economy on the weights of obese adolescent girls. Values represent mean weights under the listed conditions. Graphs 1 and 3 indicate that the two treatments interacted. No interaction is evident in graph 2, 4, or 5.

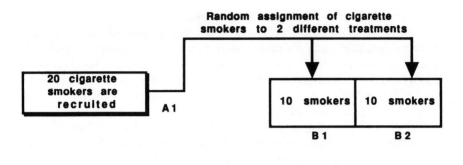

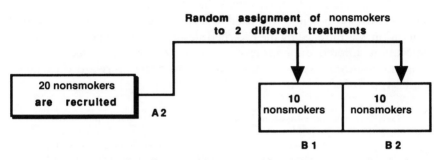

FIGURE 6.5.   Demonstration of how classification variables can be incorporated into a factorial experiment. In this example, factor A is the classification variable and factor B is the treatment (independent) variable.

to the classification variable (e.g., smokers and nonsmokers would be randomly or conveniently sampled from the population of smokers and nonsmokers). The classification variable, then, is factor A. Subjects from each level of classification are randomly assigned to the levels of the treatment (independent) variable.

Data from this type of design can provide information on factor A (classification) independently, factor B (treatment) independently, and the interaction of the two. The number of levels of the classification factor is determined by the experimental question and by the nature of the classification variable (e.g., there are only two sexes of humans, but smokers and nonsmokers can be further divided into gradients such as never smoke, smoke occasionally, or smoke daily).

Mixed Within-Subject and Between-Subjects Designs

Mixed designs combine features of within-subject and between-subjects arrangements. A study by LeSage, Makhay, DeLeon, and Poling

(1994) provides an example of a mixed design. In this study, six rats were randomly assigned to a fixed-time 60-sec (FT 60-sec) schedule of food delivery condition, and six others were randomly assigned to a massed-food condition. All subjects in each group were exposed to several doses of *d*-amphetamine and diazepam. The dependent variable was the number of fecal boli (turds) deposited in 30-min sessions. One between-subjects factor with two levels was of interest, that being the manner in which food was delivered (massed or under an FT schedule). Two within-subjects factors (*d*-amphetamine and diazepam), each with six levels (doses), also were studied. The FT 60-sec schedule was used to produce schedule-induced defecation, which was the behavior of interest in the study, and the massed-food baseline was used as a control to determine (1) whether defecation was induced by the FT schedule and (2) whether drug effects on defecation differed depending on whether or not the defecation was schedule-induced. Results of the study were straightforward (LeSage et al., 1994, p. 787):

> During vehicle control sessions [when no active drug was given], rats exposed to the FT 60-s schedule excreted a significantly higher number of fecal boli than rats exposed to massed-food sessions. *d*-Amphetamine, at doses above 0.56 mg/kg, significantly reduced defecation (boli produced) in both groups, although the magnitude of the drug's effect was larger in the group exposed to the FT 60-s schedule. For both groups, diazepam only produced a significant decrease in defecation at the highest dose (5.6 mg/kg). These results appear to be inconsistent with interpretations of adjunctive behavior that emphasize arousal or emotion as mechanisms.

Figure 6.6 shows the results obtained with diazepam. Designs similar to that used by LeSage et al. (1994) are useful whenever a researcher wants to compare the effects of a particular independent variable across conditions that cannot readily be arranged on a within-subjects basis. In the study by LeSage and colleagues, it was not practical to arrange both FT 60-sec and massed-food baselines for individual rats, because doing so on a daily basis would have resulted in overfeeding and doing so across days might have diminished the schedule-induced behavior.

In another common mixed design, subjects are randomly assigned to levels of a treatment factor, and repeated measures of the dependent variable are taken at several points in time. Thus, time can be conceptualized as a second factor. A good example of this type of design is provided by Frisch and Dickinson (1990). The authors were interested in the effects of different levels of incentive pay on work productivity. The independent variable was the incentive pay as a percentage of base pay. Frisch and Dickinson (1990, p. 13) describe their study as follows:

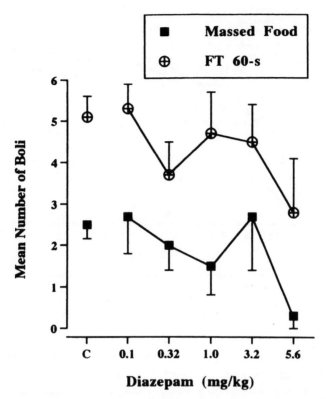

FIGURE 6.6. Mean number of fecal boli emitted in 30-min sessions by individual rats exposed to a fixed-time 60-sec (FT 60-s) schedule of food delivery or massed-food conditions. Under the FT 60-s schedule, a food pellet was presented every 60 sec regardless of the rat's behavior. Under massed-food conditions, 30 food pellets were presented together at the beginning of the experimental session. From LeSage et al., "The effects of d-amphetamine and diazepam on schedule-induced defecation in rats," *Pharmacology Biochemistry and Behavior*, Vol. 48, pp. 787–790. Copyright 1994 by Pergamon Press. Reproduced by permission.

Seventy-five college students were assigned to one of five incentive/base pay conditions: 0% (no incentives), 10%, 30%, 60%, and 100%. Subjects participated in 15 forty-five minute sessions during which they assembled parts made from bolts, nuts and washers. Subjects received a base pay amount for assembling a minimum of 50 quality parts per session and a per piece incentive for parts over 50. If subjects assembled 120 parts, the production maximum, the total amount they could earn in incentives equaled 0%, 10%, 30%, 60% or 100% of their base pay.

The results of this study are presented in Figure 6.7. The 0% incentive group clearly performed less well than the other four groups. No differ-

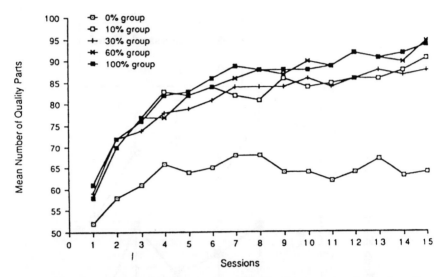

FIGURE 6.7.   Mean number of quality parts assembled in 45-min sessions by five incentive-pay groups. From C. Frisch and A. Dickinson, "Work productivity as a function of the percentage of incentives to base pay," *Journal of Organizational Behavior Management*, Vol. 11, pp. 13–33. Copyright 1990 by Haworth Press. Reproduced by permission.

ence between the 10%, 30%, 60%, and 100% groups is apparent. This design allowed the experimenters to investigate the effects of different levels of incentive pay and also to observe the changes in those effects over time.

The design used by Frisch and Dickinson (1990) was mixed in that each subject was randomly assigned to a single treatment level (pay conditions), which is a between-subjects design feature. Many kinds of mixed designs can be envisioned. Such designs are useful whenever repeated measures of behavior across time are desired and it is impossible or impractical to expose each subject to every condition of interest. When elements of within-subject and between-subjects designs are combined in the comparison of treatments, it is potentially possible to (1) isolate each subject's response to treatment, (2) summarize the responses of groups of individuals to treatment, and (3) compare the overall effects of two or more treatments, even when those treatments produce irreversible effects.

Nonexperimental Methods

Practical or ethical constraints sometimes make it difficult or impossible to conduct an actual experiment, in which the value of an independent variable is manipulated and its effects on a dependent variable are observed directly. In such cases, nonexperimental methods may be used to gain meaningful information about behavior and the variables that influence it. Nonexperimental investigations serve to describe relations among variables as they exist in the natural environment, but it is generally accepted that they are not adequate for supporting strong assertions about functional (or causal) relations. Therefore, their role in a science of behavior is limited. Three general types of nonexperimental investigations (archival, survey, and observational), classified according to the type of data collected, are discussed in the following sections.

Archival Studies

Archival studies make use of existing documentation as a data source. Usable information can be obtained from clinical records, quarterly reports, census banks, or any other permanent records kept by clinics, hospitals, businesses, governments, or other organizations. Such sources sometimes provide meaningful information about behavior, and perusing data from such sources may suggest how behavior was affected by a particular environmental event.

For example, Lavelle, Hovell, West, and Wahlgren (1992) used archival data in a study of child protection and law enforcement, these being archival records of the number of tickets issued by police officers for nonuse of child safety seats. They examined police records from two communities, one in which a program to target child safety-seat use was implemented and one in which there was no program. In the intervention community, officers were instructed on the importance of safety-seat use and were told that when they issued tickets for safety-seat infractions they were also to give drivers coupons redeemable for safety seats and waivers for fines (good if the recipient attended a special safety class). Three years of baseline data were examined for the intervention community, followed by 6 months under the safety-seat program and 12 subsequent months after withdrawal of the program. The other community, in which conditions did not change systematically across time, was monitored over the same period (i.e., 54 months). The archival data, presented graphically in Figure 6.8, show that officers in the intervention community had a much higher rate of issuing tickets for nonuse of child safety seats during the intervention period compared to baseline levels.

Levels of ticket writing dropped during the reversal phase, but were maintained at a higher rate than during the initial baseline period. Officers in the comparison community maintained low levels of citations for nonuse of safety seats throughout the 54 months. These results certainly

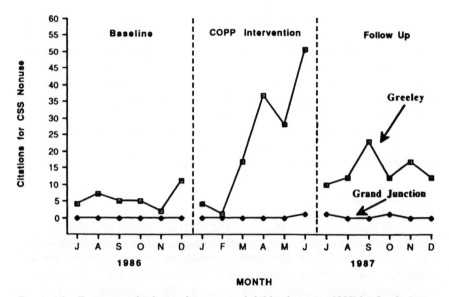

Figure 6.8. Frequency of ticketing for nonuse of child safety seats (CSS) by Greely (intervention site) and Grand Junction (control site) officers for the 6-month baseline, intervention, and follow-up periods. Officer training began at the beginning of the intervention (COPP) phase for the Greely group only and was completed within 8 weeks. From Lavelle et al., "Promoting law enforcement for child protection: A community analysis," *Journal of Applied Behavior Analysis*, Vol. 25, pp. 885–892. Copyright 1992 by the Society for the Experimental Analysis of Behavior. Reproduced by permission.

suggest, as Lavelle et al. (1992) pointed out, that the program affected ticket-writing behavior. This conclusion is reasonable, even though the investigators did not manipulate an independent variable or even plan a prospective study. Rather, they took advantage of an existing body of information from which they drew their conclusions.

Although archival data can support meaningful conclusions, one should be aware of the limitations of such data. First, the content included in archival records should always be held as suspect. What goes into a permanent record is often as much a function of who creates the document as of what actually occurred. Second, the completeness of records as they are maintained over time should be questioned. Documents are lost, thrown away, or misfiled, or disappear otherwise, sometimes selectively (e.g., those that contain information that makes an organization look bad may be especially likely to turn up missing), making archival records incomplete, and potentially biased, sources of data. Third, errors in recording and transcribing can occur. Therefore, the accuracy of information contained in archival records should be verified whenever possible.

Surveys

A second type of nonexperimental investigation, the *survey*, is an organized way of obtaining information directly from people by asking them questions concerning their current and historical practices, opinions, and demographic characteristics. Either open-ended or closed questions can be used in survey research, and these questions can be asked in interviews or in questionnaires. Open-ended questions require respondents to supply answers in their own words. Closed questions limit response options through the use of checklists (e.g., "check the foods that you eat weekly"), Likert-type scales (e.g., "Circle a number from 1 to 10 indicating how satisfied you are with your job, with 1 indicating totally dissatisfied and 10 totally satisfied"), ranking (e.g., "List your top three choices for the president of ABA"), or yes/no options (e.g., "Do you ever read JEAB?").

An example of the use of survey methods is provided by Poling, Picker, Grossett, Hall-Johnson, and Holbrook (1981), who sent a questionnaire to the editorial staff of the *Journal of the Experimental Analysis of Behavior* (JEAB) and the *Journal of Applied Behavior* (JABA) in an attempt to gain information about the relationship between the experimental and applied areas of behavior analysis The responses of JEAB and JABA editors to the questionnaire are shown in Table 6.1. Poling et al. (1981, p. 99) summarized these data as follows:

> Due caution is called for in interpreting these data. Forced-choice questionnaires sample at best a limited behavioral repertoire, and are for good reason not favored assessment tools in behavioral psychology. This notwithstanding, responses to the questionnaire do parallel citation data [also presented in the article] in suggesting that the experimental analysis of behavior, represented by JEAB, and applied behavior analysis, represented by JABA, are largely insulated from each other. Few of the individuals we queried had published in, regularly read, or benefit from both journals; most indicated the fields are growing apart. To some extent, this may reflect our sample. The editorial staffs of JEAB and JABA may be particularly "hard-nosed" with respect to their commitment to the experimental analysis of behavior and applied behavior analysis, respectively, and other behaviorists may be more influenced by, and active in, both fields. Nevertheless, the editorial staffs of JEAB and JABA are especially successful scientists and scholars. They certainly are in a position to influence publication policies, and the behavior of students and colleagues as well. Thus, their evaluation of the relationship between the experimental analysis of behavior and applied behavior analysis is not inconsequential.

TABLE 6.1. Summary of the Results of a Questionnaire Mailed to Members of the Editorial Boards of *JEAB* and *JABA*[a]

1. At some time, the person subscribed to both *JEAB* and *JABA*.

	JEAB Editors			*JABA* Editors	
Yes	No	Other	Yes	No	Other
13 (48%)	14 (52%)	—	19 (53%)	17 (47%)	—

2. In 1980, the person subscribed to both *JEAB* and *JABA*.

	JEAB Editors			*JABA* Editors	
Yes	No	Other	Yes	No	Other
8 (30%)	19 (70%)	—	6 (17%)	30 (83%)	—

3. The person has published in both *JEAB* and *JABA*.

	JEAB Editors			*JABA* Editors	
Yes	No	Other	Yes	No	Other
1 (4%)	26 (96%)	—	4 (11%)	32 (89%)	—

4. Do you currently read *JEAB* on a regular basis?

	JEAB Editors			*JABA* Editors	
Yes	No	Other	Yes	No	Other
26 (96%)	1 (4%)	—	4 (11%)	31 (86%)	1 (3%)

5. Do you currently read *JABA* on a regular basis?

	JEAB Editors			*JABA* Editors	
Yes	No	Other	Yes	No	Other
7 (26%)	20 (74%)	—	34 (94%)	2 (6%)	—

6. Is *JEAB* of significant value to you in your current research efforts?

	JEAB Editors			*JABA* Editors	
Yes	No	Other	Yes	No	Other
25 (93%)	2 (7%)	—	7 (19%)	26 (72%)	3 (9%)

7. Is *JABA* of significant value to you in your current research efforts?

	JEAB Editors			*JABA* Editors	
Yes	No	Other	Yes	No	Other
7 (26%)	20 (74%)	—	30 (83%)	5 (14%)	1 (3%)

8. Across time, research articles published in *JEAB* have _____ in value to applied behavior analysts.

	JEAB Editors				*JABA* Editors		
De-creased	In-creased	Not Changed	Other	De-creased	In-creased	Not Changed	Other
3 (11%)	4 (15%)	15 (55%)	5 (19%)	22 (61%)	2 (5%)	6 (17%)	6 (17%)

9. Do you feel that the field of applied behavior analysis and the experimental analysis of behavior have become increasingly separate from one another?

	JEAB Editors			*JABA* Editors	
Yes	No	Other	Yes	No	Other
20 (74%)	3 (11%)	4 (15%)	30 (83%)	5 (14%)	1 (3%)

10. If your answer to question 9 was "yes," do you feel that the separation of the fields is harmful to behavioral psychology?

	JEAB Editors			*JABA* Editors	
Yes	No	Other	Yes	No	Other
11 (55%)	8 (40%)	1 (5%)	16 (53%)	13 (43%)	1 (4%)

[a]From Poling et al. (1981). The actual questionnaire is available in their article.

Survey methods are useful for obtaining anonymous information from large and widely distributed groups of people, and this is their primary advantage. Because survey methods can allow respondents anonymity, these methods are often used to gain information about troublesome behaviors for which interventions are needed. For example, nationwide surveys of drug use indicate that cigarette smoking peaked in the United Stats in 1963 and has declined steadily since 1963 (Maisto, Galizio, & Conners, 1995). Nonetheless, nearly 30% of respondents in the 1992 survey reported cigarette use in the past month (U.S. Department of Health and Human Service, 1993). Hence, further attention to stop-smoking and don't-start-smoking programs is warranted.

Although survey methods are relatively easy to use, the information reported in a survey may not be an accurate reflection of the actual behaviors of interest. This possible inaccuracy is a serious disadvantage. Moreover, how people respond to a survey depends, in part, on how the survey is constructed. Detailed information about the use of survey methods is provided in several sources (e.g., Ary, Jacobs, & Razavieh, 1990; Dillman, 1978).

Observational Studies

In an *observational study*, there is no experimental manipulation, and data are simply recorded under the conditions that occur in the natural environment. These conditions may differ substantially, however, and observational studies are often mistaken for true experiments. They are not, and they provide less compelling demonstrations of relations between variables than can be provided by true experiments.

In a good example of an observational study, Cohen-Mansfield, Marx, and Werner (1992) examined the relationship between time use and behavior problems in a nursing home. In this study, elderly nursing home residents were observed as they functioned from day to day in the residential environment. The authors reviewed current literature and concluded that little is known about the etiology and course of agitation in elderly persons. They hypothesized, on the basis of clinical observations, that inactivity plays an important role in agitation in elderly residents of nursing homes and addressed two specific questions in their study: How do agitated residents spend their waking hours? Are the type and amount of agitation associated with time use?

To answer these questions, they observed 24 residents of a 550-bed long-term-care nursing home. Their data, shown in Table 6.2, indicated an inverse relationship between agitation and structured activity level; that is,

Table 6.2. Means of Activity–Behavior Combinations for 23 Agitated Residents of a Nursing Home for Severely Cognitively Impaired Individuals[a]

Behavior	Activities[b]				
	None	ADL	Relocate	Social	Structured
Repetitious mannerisms	2.60	1.16	0.61	1.77	0.96
Strange movements	0.40	0.11	0.08	0.07	0.06
Picking at things	1.02	0.47	0.84	0.83	0.45
Strange noises	1.08	0.67	0.66	0.79	0.25
Constant requests for attention	1.96	1.81	1.78	1.98	0.85
Pacing	0.38	0.18	7.93	1.55	0.12
Aggressive behaviors	0.11	0.09	0.13	0.13	0.09

[a]From Cohen-Mansfield et al. (1992). Copyright 1992 by the Southern Gerontological Society. Reproduced by permission. Each number is the mean of approximately 1000 3-min observations collected across 3 months. The higher the score, the higher the level of the indicated behavior.
[b]Activities: (None) no activity (unoccupied); (ADL) activities of daily living (eating, toileting or bathing, grooming, and getting up or transferring); (Relocate) relocating from one place to another); (Social) social interactions (e.g., visiting); (Structured) structured activities (e.g., organized activities, watching television, play).

a greater frequency of agitated behaviors was observed when residents were less occupied, except when they were being relocated.

The results obtained by Cohen-Mansfield et al. (1992) do not merit the interpretation that being unoccupied caused residents of the nursing home to become agitated. Nor can one conclude that being agitated caused residents to avoid being occupied in structured activities. While one of these interpretations may reflect the true state of affairs, there is no empirical support for either. It could be the case that some third factor caused both agitation and avoidance of activities; perhaps a particular drug therapy is responsible. Or perhaps residents with poor communication skills become agitated and unoccupied. The point is, an interesting relationship was described by observing behavior without manipulating an independent variable, but we do not know the exact nature of that relationship. Further study, using experimental methods, would be necessary to clarify it.

Nonexperimental investigations may contribute useful information to a science of behavior. They require careful forethought and the same meticulous concern for detail as do good experimental investigations. Nonexperimental studies do not support causal interpretations, but their findings may describe naturally occurring relations that are of interest in themselves, or that merit further inspection via experimental analysis.

Summary and Conclusions

In between-subjects designs, levels of one or more independent variables are varied, and comparisons are made across groups of subjects. While group comparisons are useful for studying behavioral phenomena in certain contexts, their applications in behavior analysis are limited. Barlow and Hersen (1984) classify the limitations of between-groups approaches under five headings: ethical objections, practical problems, using group averages for comparison, generality of findings, and intersubject variability. Ethical objections are most often raised in relation to withholding of treatments from control group subjects in clinical studies. Barlow and Hersen (1984, p. 14) contend that this notion is "based on the assumption that the therapeutic intervention, in fact, works, in which case there would be little need to test it at all." One alternative to the control group approach is to use a comparison group comprised of subjects exposed to a treatment that has been empirically demonstrated to be effective. The effects of the experimental manipulation are then contrasted with the known effects of the comparison treatment.

Practical problems arise in procuring large numbers of subjects possessing the characteristics of interest for a particular study. These problems are particularly troublesome when specific clinical populations are of interest.

Although this is a data analysis issue, not a design issue, basing conclusions about the effectiveness of an intervention on group averages is common in between-subjects experiments, and troublesome. Sidman (1960) argued that a mean score is rarely representative of all subjects in a population. He suggested that published research should always include some statement of the number of subjects who actually fell at the mean level on the dependent measure. In his words (Sidman, 1960, p. 48), "If the group were sufficiently large, we would then be able to state that, for example, 30 per cent of the subjects will yield a mean value, y, of the behavioral measure. This would be a true statement of the degree of representativeness of the group mean." The information that Sidman suggests be provided is rarely provided, however.

Mean data alone provide no indication as to how individual subjects respond to an intervention. Barlow and Hersen (1984) note that group averages have the undesirable characteristics of obscuring individual clinical outcome. This problem is worsened when some subjects in a group improve under treatment conditions and some subjects worsen. In light of these difficulties with between-groups comparisons, it is good practice in presenting results to include individual scores under all treatment (and control) conditions, preferably as a clear graph. These data provide stimuli in addition to group means to which consumers can respond in interpreting the results of an experiment.

Both between-groups and within-subjects comparisons present the experimenter with problems concerning the generality of results. With respect to between-groups comparisons, obtained results can be said to generalize to the population from which the subjects were drawn. But what of generality to individual members of the population? Sidman (1960, p. 49) points out that the behavior of the individual member is "an orderly function of a large number of so-called independent variables." He goes on to say that because individual functions are not obtained in the group assay, it is impossible to bring individual differences to light. Thus, it is also impossible to extend findings of such research to specific individual members of a population.

The final weakness discussed by Barlow and Hersen (1984), intersubject variability, is often the source of weak results garnered in between-subjects studies. When the behavior of subjects is measured at one point or a few points in time and group data are considered, the response of specific individual subjects undergoing treatment is ignored. When some subjects improve and others deteriorate, the average gain may be small or nonexistent. The typical behavior analyst is interested, however, in the course of individual changes over time under some treatment manipulation. Such effects can be detected only through the use of either within-subject or mixed designs. These designs are useful in many situations, and they play a fundamental role in the analysis of behavior.

Despite the difficulties associated with the use of between-subjects designs in behavior analysis, these designs have legitimate applications. Such designs may be appropriate for large-scale studies, for comparing different treatments, and for investigating the interaction between independent variables. Thematic research can make use of between-subjects and within-subject designs to answer different questions at different stages. For example, an investigator may begin a line of research with a within-subject design and discover an important functional relation. A between-subjects design can be subsequently employed to ascertain if that relationship holds across different populations of subjects. Alternatively, a research project can begin with a large-scale between-subjects investigation to demonstrate the existence of a relation between an independent variable and a target behavior. The researcher might then arrange a within-subject design to determine whether the relation holds over time with the same subjects.

Within-subject and between-subjects designs provide fundamentally different kinds of information and are not usually interchangeable. Both are capable, however, of demonstrating functional relations, and both have a role in a science of behavior.

Nonexperimental methods, in contrast, can suggest functional relations, but not confirm them. Archival, survey, and observational methods have played a limited role in behavior analysis, but do provide a means of obtaining information that, for practical or ethical reasons, cannot be garnered through experimental methods. Further, nonexperimental methods can be gainfully employed in the early stages of planning experimental investigations, particularly in applied behavior analysis. Aspiring behavior analysts should be familiar with these methods, but use them sparingly, and only for good reason.

CHAPTER 7

Graphic Analysis of Data

Scientific data are important only insofar as they influence the behavior of the researcher and of other people. As Michael (1974, p. 647) points out:

> The observations [data] resulting from scientific experiments are stimuli that hopefully affect the scientist and his [or her] colleagues by producing better practical behavior, more sophisticated follow-up experiments, or better verbal behavior regarding the subject matter. These stimuli, however, may not result in any effective reaction, a fairly common reason being their complexity. Repeated observation of the same experimental condition, for example, may give rise to a set of numbers, all differing considerably from one another. This situation has occurred quite often and methods have been discovered for simplifying it to some degree. Some of the methods generate two-dimensional visual stimuli where the values of each dimension stand in a point-to-point relation to some feature of the data; a frequency polygon is such a stimulus. Another stimulus-simplifying technique results in a smaller set of numbers, each related to some important characteristic of the larger set, such as the mean and range of the raw data. Using the term "judgment" to refer to any of the various kinds of reactions that a scientist could make to the data of his [or her] experiment, it is useful to refer to these stimulus-simplifying techniques and their products as "judgement aids."

Graphs that depict experimental data are judgement aids; they make it easier to respond to the results of an experiment. Consider the hypothetical data presented in Table 7.1.

Given these data in tabular format, would you say a treatment effect has been demonstrated? How confident are you? How long did it take to decide? Now evaluate the same set of data as presented in Figure 7.1. It is quickly apparent from the data in graphic format that levels of the target behavior were consistently lower when treatment was present than when it was absent. Thus, one can be reasonably confident that a treatment effect occurred.

Graphic (visual) analysis is ubiquitous in behavior analysis. As evidence of this omnipresence, McEwan (1995) determined the average number of graphs per page in every research article published in 1992 in the

TABLE 7.1. Hypothetical Data Set

Baseline	Treatment	Baseline
32, 45, 29, 30, 35, 42, 39, 48, 39, 43, 40, 37	20, 15, 25, 32, 28, 34, 22, 21, 24, 19, 23, 20	35, 31, 40, 44, 37, 42, 46, 39, 36, 41, 47, 32

two key behavior analytic journals, the *Journal of the Experimental Analysis of Behavior* (JEAB) and the *Journal of Applied Behavior Analysis* (JABA). On average, there were 1.5 graphs per page in JEAB articles and 0.75 graph per page in JABA articles. McEwan also reported that, on average, 1 in 5 articles published in JEAB used inferential statistics to analyze data, whereas 1 in 15 articles published in JABA did so. Every article that presented a statistical analysis also presented one or more graphs. Clearly, in behavior analysis, graphs are used as judgment aids far more often than are inferential statistics. This proportion is unsurprising; the advantages of graphic data analysis are well documented (Cooper, Heron, & Heward, 1987; Johnston & Pennypacker, 1993b; Michael, 1974; Parsonson & Baer,

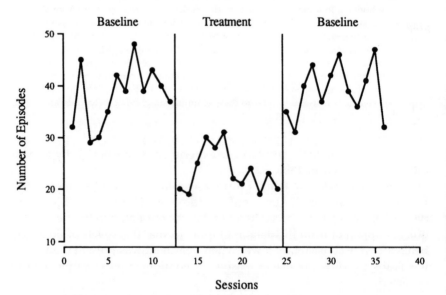

FIGURE 7.1.   Hypothetical example of a line graph used to portray data on the effects of an intervention on the number of episodes of a target behavior. In this hypothetical study, a withdrawal design was used to examine the effects of the intervention.

1986; Sidman, 1960; Tawney & Gast, 1984). Three advantages of particular importance are discussed below.

Advantages of Graphic Analysis

One advantage of graphic analysis is its practicality. Graphic analysis is often quick and effective and is relatively easy to learn. Moreover, graphic analysis can be applied to a wide range of data. As Parsonson and Baer (1986, p. 157) note, graphs provide a "comprehensive, yet simple, means of recording, storing, representing, communicating, and above all, analyzing behavioral data."

A second advantage of graphic analysis is that if data are graphed and evaluated after each observational period (e.g., day or experimental session), the researcher can change experimental conditions as dictated by the behavior of individual subjects. Assume, for example, that a researcher using a probe design to evaluate the effects of a particular drug on learning in monkeys initially planned to evaluate doses of 0.01, 0.1, and 1 mg/kg, with the doses given in random order. Monitoring incoming data, the researcher discovers that none of these doses is behaviorally active. Given this discovery, it is reasonable to add higher doses to the schedule.

When using within-subject experimental designs, *it is good practice to graph data for each subject after each observational period and to determine whether the incoming data indicate that a change of conditions is necessary.* This practice obviates many potential problems and substantially increases the likelihood of producing meaningful results.

A third advantage, and one that is especially important from an applied perspective, is that many people consider graphic analysis to be a conservative method for evaluating the efficacy of an intervention (Cooper et al., 1987; Kazdin, 1982; Parsonson & Baer, 1986). According to these people, the scientific community will be convinced of the efficacy of an intervention only when its effects are large enough to be obvious in a graphic portrayal of the data. Interventions that produce weak or variable effects are less likely to be considered important. Consequently, researchers using graphic analysis are less likely to commit Type I errors (discussed in Chapter 8) than those who rely on inferential statistics (Parsonson & Baer, 1986). That is, they are less likely to conclude erroneously that an intervention was effective. Therefore, graphic analysis can serve to filter out weak interventions in favor of powerful interventions and thereby help to develop a very powerful behavior-change technology (Baer, 1977).

Interestingly, not everyone agrees that graphic analysis is especially conservative. Huitema (1986b) points out that there is no empirical support for the notion that graphic analysis is insensitive to small effects. As Huitema (1986, p. 215) indicates, "The amount of data and the variability within and between conditions largely determine the identification of effects. This is true for single- subject and group designs and for visual and statistical analyses."

Disadvantages of Graphic Analysis

As a rule, graphic analysis is superior to statistical analysis and is to be preferred in behavior analytic research. Nonetheless, graphic analysis does have potential disadvantages. One is that it may be difficult or impossible to apply graphic analysis when many data, collected under a complex arrangement of conditions, are of interest. In such cases, which are more likely to occur in areas such as epidemiology (the science that studies the variables that determine the frequency and distribution of diseases) than in behavior analysis, it may be necessary to take recourse to statistical analyses to make sense of the findings.

A second disadvantage is that some scientific communities are unwilling to accept conclusions based on graphic analysis alone. Publishing papers in journals controlled by these communities requires researchers to use inferential statistics, although nothing prevents one from presenting and discussing detailed graphs as well as summary statistics.

A third alleged disadvantage is that there are no formal rules for conducting graphic analyses, and in some cases, people may disagree on how they evaluate particular data sets (e.g., De Prospero & Cohen, 1979; Gottman & Glass, 1978). There are, in contrast, formal rules for calculating inferential statistics; hence, all persons using a given test to analyze a particular set of data should agree as to whether or not behavior differed across conditions.

This disadvantage is more apparent than real. There are general rules for preparing and evaluating graphs; several are presented in this chapter and elsewhere (e.g., Cooper et al., 1987; Johnston & Pennypacker, 1993b; Parsonson & Baer, 1986). Furthermore, there often is substantial agreement among professionals on whether or not a treatment effect is evident in a particular graph. Were this not so, referees could never agree on whether articles submitted to JEAB or JABA demonstrated a treatment effect and merited publication, and they often do agree. Often, but not always.

But simply that people disagree on how they evaluate the results of a study does not mean there is a problem. Contrary to popular belief, there is no surety in science. Some patterns of results are clear and strongly support obvious inferences about the relation between an independent and a dependent variable. Others are chaotic; everyone agrees that there is no evident relation between variables. Between these endpoints lie patterns of data that are to some degree ambiguous. No matter how those data are analyzed, their fundamental nature does not change. Graphic analysis does not hide ambiguity and force a consensus as to whether or not a treatment effect was observed. Instead, it allows individual researchers to evaluate for themselves the effects of an intervention and to behave accordingly. Allowing independent evaluations of an intervention's effects is, in fact, a strength of graphic analysis, not a weakness.

Constructing Line Graphs

The most commonly used graph format in behavior analytic research is the line graph, like that in Figure 7.2. Researchers in different areas of behavior analysis follow different graphing conventions, and line graphs can differ substantially in appearance. Nonetheless, all line graphs share common features. Several authors have provided detailed discussions on how to construct and interpret graphs (e.g., Bowen, 1992; Johnston & Pennypacker, 1993b; Tufte, 1983). Exceptionally clear and straightforward coverage is provided by Cooper et al. (1987), and our discussion of the construction of line graphs borrows heavily from their work.

In general, a line graph consists of a horizontal axis, a vertical axis, phase-change lines, data points, data paths, labels, a figure legend, and a figure caption. Miscellaneous other symbols may also appear. Each of these components is number-labeled in Figure 7.2 and numbered correspondingly in the following discussion.

1. *Horizontal axis (x-axis or abscissa).* The horizontal axis is a horizontal line that characteristically depicts the passage of time and the presence, absence, or value of the independent variable. It is usually marked off in equal intervals, each representing an equivalent period of time. Chronology is represented by numbering units of time (e.g., minutes, hours, days) or consecutive observational periods (e.g., trials, sessions) from left to right. When the horizontal axis is used to scale specific values of the independent variable, the values are represented from the lowest on the left end of the axis to the highest on the right end. The length of the horizontal axis is determined by the amount of data to be graphed. The space between intervals should be sufficient that data points do not crowd together or overlap. Data points should be spaced far enough apart to allow one to discern patterns of performance easily. When there are several intervals on the horizontal axis (e.g., days, sessions), only some of the

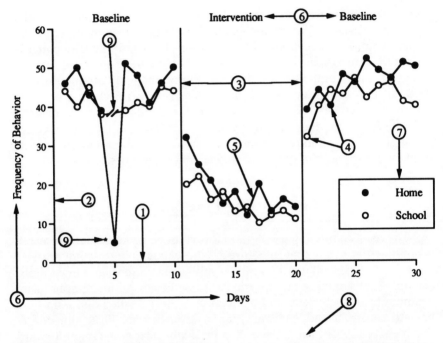

FIGURE 7.2.   Hypothetical example of a line graph. The major parts of a simple line graph are:
(1) horizontal axis, (2) vertical axis, (3) phase-change lines, (4) data points, (5) data path, (6)
labels, (7) figure legend, (8) figure caption, and (9) miscellaneous symbols [line break (//)
and asterisk (*)].

intervals should be numbered (e.g., every 2nd, 5th, or 10th interval) to
avoid unnecessary clutter. When multiple graphs are presented one above
the other in the same figure, it is necessary to number and label only the
horizontal axis on the bottom graph, provided the horizontal axes of all
graphs above it are scaled the same (see Figure 7.3). There should be tick
marks on the horizontal axes of all the graphs, however, and they should
be aligned to facilitate comparisons across graphs. *Tick marks* are the short
lines that extend downward from the horizontal axis and leftward from
the vertical axis to demarcate the units of measurement.

Another frequently used convention, also illustrated in Figure 7.3, is
that of representing standard deviations by vertical brackets at the data
points.

2. *Vertical axis (y-axis or ordinate).* The vertical axis is a vertical line
drawn upward from the left end of the horizontal axis. The point at which
the vertical and horizontal axes intersect is called the *origin.* The vertical

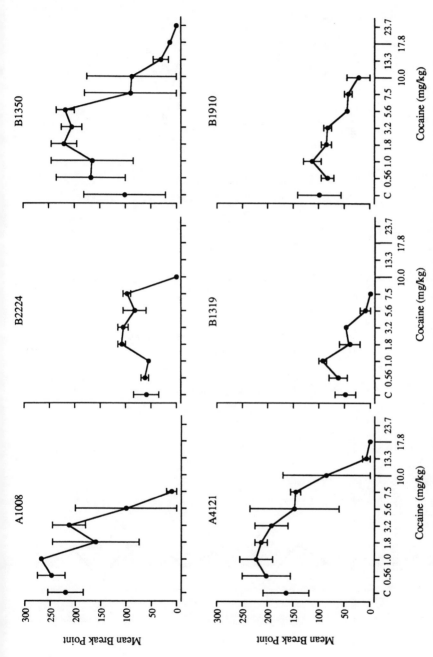

FIGURE 7.3. Example of a figure displaying multiple graphs. Only the vertical axes of the far left graphs and the horizontal axes of the bottom graphs are labeled. From C. A. Jones, LeSage, Sundby, and Poling (1995). Copyright 1995 by Pergamon Press. Reproduced by permission.

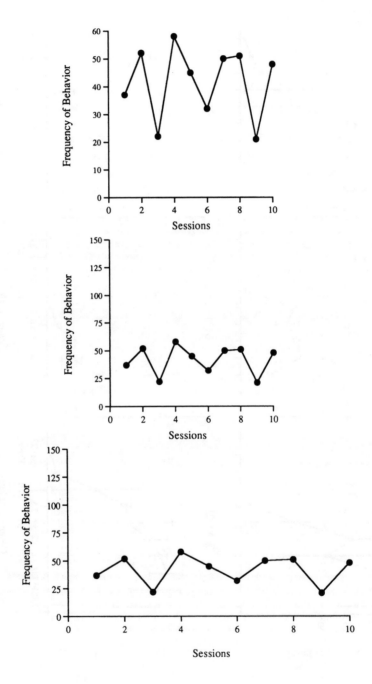

axis depicts the level of the dependent variable and is always scaled along some quantifiable dimension of behavior. It is usually marked off in equal intervals, each representing an equal amount of behavior. The values of the dependent variable are represented from the lowest (usually 0) at the origin to the highest at the top end. So long as the values portrayed do not change, the length of the vertical axis determines the apparent variability of the data displayed in the graph; the shorter the axis, the smaller the apparent variability. A short vertical axis or a vertical axis with a maximum value that far exceeds the highest level of the dependent variable can mislead the reader by obscuring otherwise large variability (see Figure 7.4). Most authors suggest that the proportionate length of the vertical axis to that of the horizontal axis should be about 2:3 (e.g., Bowen, 1992; Cooper et al., 1987). There appears to be no empirical reason for selecting this proportion, and it often varies as a function of the amount of data to be displayed and the number of graphs in a figure. As the amount of data and the number of graphs to be displayed increase, the length of the vertical axis characteristically decreases. When graphs are displayed side by side in the same figure, it is necessary to number and label only the vertical axis on the far left graph, provided the vertical axes of all graphs to the right of it are scaled the same (see Figure 7.3). There should be tick marks on the vertical axes of all the graphs, and they should be aligned horizontally to allow comparison of levels of performance between graphs.

   3. *Phase-change lines.* Phase-change lines are solid vertical lines drawn upward from the horizontal axis that demarcate the points in time at which the independent variable was manipulated (e.g., treatment was applied or withdrawn). The lines are placed between the last data point of one phase and the first data point of the subsequent phase. Phase-change lines are never drawn through data points. When an intervention phase has subphases (e.g., in a changing-criterion design), these subphases are demarcated by dashed vertical lines (see Figure 7.5). When portraying data from a multiple-baseline design, the phase-change line is extended to just below the horizontal axis of the upper graph, angled 90° to the right, then angled 90° downward at the point at which the phase-change line extends down to the horizontal axis of the lower graph (see Figure 7.6).

   4. *Data points.* Data points are symbols that represent (1) the value of

FIGURE 7.4. Graphs of hypothetical data showing changes in the apparent amount of variability as one manipulates the range of the vertical axis and the length of the horizontal axis. The middle graph shows the decrease in apparent variability as one increases the range of the vertical axis (relative to the range of the vertical axis in the top graph). The bottom graph shows a further decrease in apparent variability as one lengthens the horizontal axis (relative to the length of the horizontal axis in the top graph) and increases the range of the vertical axis.

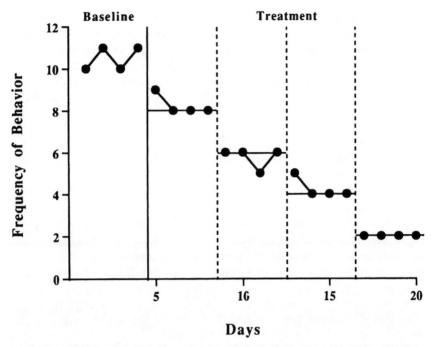

FIGURE 7.5.   Hypothetical data obtained from a changing-criterion design. The major phase change between the baseline condition and the treatment condition is marked by a solid phase-change line. Changes between subphases of the treatment are marked by dashed phase-change lines.

the dependent variable and (2) the time at which and the experimental condition under which that value was obtained. For example, in Figure 7.2, each data point represents the number of times per day (frequency) the target behavior occurred under baseline and intervention conditions. The symbols should be large enough for the reader to see them easily, but small enough to avoid cluttering the display. It is convention to use solid black dots for symbols when only one data set is plotted. When more then one data set is plotted, a different geometric symbol should be used for each data set. Generally, when two data sets are plotted, a solid symbol is used for one data set and an open symbol for the other, as shown in Figure 7.2. *The first data point is never plotted directly on the vertical axis, and subsequent points are never plotted directly on phase-change lines.*

5. *Data path.* The data path is a line that connects the data points and represents the overall relationship between the independent and dependent variables. When multiple data sets are plotted, a different style of line

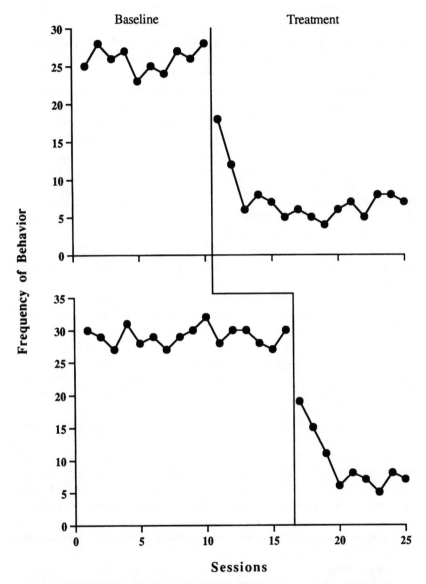

FIGURE 7.6. Hypothetical data obtained from a multiple-baseline design showing the stag-gered phase-change line across baselines (i.e., graphs).

may be used for each data set. Different data-point symbols, however, are usually sufficient to differentiate multiple data sets. In some cases, a few data points will represent extreme values of the dependent variable (i.e., will lie well beyond the range of the other data points). In such cases, the extreme data points can be placed just above the end of the vertical axis with an interrupted data path going to and from that data point. The value is printed next to the data point representing that value. (Another use of the line break is illustrated in Figure 7.2 and discussed in paragraph 9 below.) *Data points on opposite sides of a phase-change line should not be connected.* As Parsonson and Baer (1986) note, when a data path crosses a phase-change line, it can underemphasize a change in performance across phases.

6. *Labels.* Labels are single words or brief phrases that identify the horizontal and vertical axes and each phase portrayed in the graph. They are extremely important in making clear to the reader what the data in a graph represent. An axis label identifies the dimension along which the axis is scaled. The specific units of each scale should be identified in the label. For example, in Figure 7.2, the horizontal axis is scaled along the dimension of time, measured in units of days. *Labels that identify phases of the experiment should be placed at the top of the graph parallel to the horizontal axis.* An error commonly made by novices is to place these labels under the horizontal axis. All labels should be unambiguous, and uncommon abbreviations should be avoided.

7. *Figure legend.* The figure legend aids the reader in identifying what each symbol or data path represents. When different data sets are represented by different plot symbols, the legend should contain one of each symbol and a word or brief phrase describing what the symbol represents. In Figure 7.2, the solid and open circles are identified as representing data obtained at home and at school, respectively. When different line styles are used to identify different data sets, the legend should contain a short sample of each line style and a word or phrase describing what each line style represents. The legend can be placed in any open space within or outside the graph frame. It should not overlie any other part of the graph (e.g., axes, data paths) and should be framed to help the reader distinguish it from the rest of the graph. Without a frame, the legend can make a graph look cluttered. A figure legend is not needed when only one data set is plotted. In this case, a figure caption is sufficient to provide all the descriptive information.

8. *Figure caption.* The figure caption is an informative but concise statement describing what the graph portrays. The figure caption, axes, phase-change lines, and labels together show the reader the relationship between the independent and dependent variables. At a minimum, the

caption should state the dimension along which each axis is scaled, the phases of the study, and, in the case of multiple data sets, what each plot symbol or line style represents. It may seem redundant to identify plot symbols or line styles in both the figure legend and the figure caption. Realize, however, that the figure legend is a visual aid to the reader and obviates repeated reference to the figure caption as the reader examines the graph. Moreover, the figure caption can provide more detailed information than the figure legend. The figure caption should explain any excessively deviant data points, confusing or potentially misleading features of the graph, and miscellaneous symbols.

9. *Miscellaneous symbols.* Symbols are often included in graphs to identify unusual events. For example, subjects may become ill, fail to attend a session, or perform at an unusually high or low level. An asterisk (*) or a line break (/ /) is often used to identify such an event. For example, the asterisk in Figure 7.2 indicates that the unusually low level of behavior at home on day 5 was likely due to illness. The line break in the data path between days 4 and 6 indicates that no data were obtained at school on day 5 because the subject did not attend school. Line breaks are also used to break the vertical axis when some deviant data points are present in the data set or when scaling changes are made. Line breaks are used to break the horizontal axis when an interval represents a passage of time longer than that of the other intervals or when changes in scaling are made.

Bar Graphs

Bar graphs also are used to depict behavioral data. Bar graphs are constructed in the same way as line graphs, except that values of the dependent variable are represented by bars instead of data points. Because the bars take up considerable space along the horizontal axis, bar graphs are inferior to line graphs when many observations are of interest and the researcher is unwilling to summarize them (e.g., as means). Summarizing data, regardless of whether the data are presented in a bar graph, a line graph, or a table, may make it easier to respond to those data. The risk is that the conclusions supported by summary data may not be the same as those supported by raw (untransformed) data. For example, Figure 7.7 shows a bar graph (top) that summarizes the data obtained under a baseline and treatment condition. This graph, which shows mean levels of behavior across the two conditions, provides a simple and easily interpretable depiction of the findings. One can readily discern changes in performance by comparing the height of the bars, and it appears that the mean frequency of behavior increased as a function of presenting the treatment.

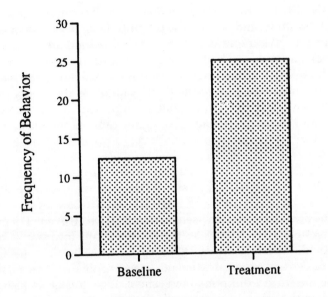

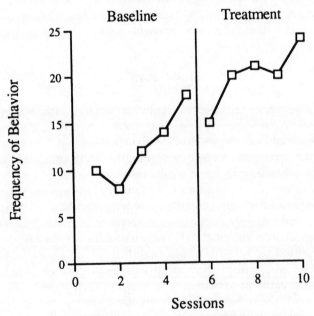

FIGURE 7.7. Two graphs of the same hypothetical data set. The bar graph at top, which shows means, obscures the trends that are evident in the line graph at bottom, which shows raw data.

This conclusion would not be supported by the raw data, which are shown in the line graph in Figure 7.7 (bottom). Given this presentation of the data, it is perfectly reasonable to conclude that the frequency of behavior may have continued to show an increasing trend in the absence of treatment (i.e., that the treatment was not responsible for the increase in behavior).

Cumulative Records

Parsonson and Baer (1986) assert that the graphic analysis of data in behavior analytic research grew out of B. F. Skinner's use of the cumulative record as a primary means of analyzing the performance of nonhuman subjects. Cumulative records (discussed in Chapter 4) are graphic representations of raw data. They provide an exceptionally precise measure of behavior, sensitive enough to reveal the subtlest effects of an independent variable, and are unexcelled in allowing for fine-grained analysis of changes in behavior during the course of an observational period (e.g., experimental session). Examining cumulative records during the course of the sessions that produce them, or shortly thereafter, is an established strategy for monitoring experiments, and it serves to reveal both potential problems and interesting but unanticipated effects of an independent variable. Many veteran researchers in the basic areas of behavior analysis make use of the strategy, and we recommend it for all laboratory researchers.

Cumulative records are rarely used in applied research (Cooper et al., 1987), due largely to the kinds of experimental questions of interest. Even in basic research areas, the popularity of cumulative records for portraying findings has declined, at least in published articles (Poling, 1979). This decline is perhaps unfortunate, because in some cases, carefully selected cumulative records can provide better depictions of performance than any alternative. Consider, for example, a pattern of responding common in animals with extensive histories of responding under relatively long fixed-interval (FI) schedules of reinforcement. In such animals, little or no responding occurs for a substantial period after a reinforcer is delivered. This period frequently lasts for roughly two thirds of the FI (e.g., 80 sec under an FI 120-sec schedule). When responding starts, it starts gradually and slowly increases until a high rate is reached at the end of the interval, when reinforcement occurs. This pattern, which is somewhat cumbersome to describe verbally or to envision on the basis of a verbal description, is readily apparent in a cumulative record, where it appears as an FI scallop (see Figure 7.8).

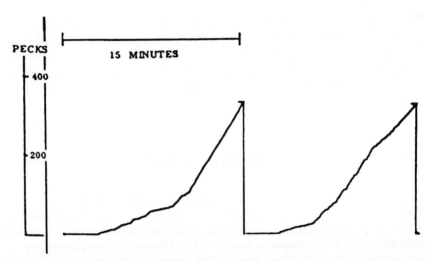

Figure 7.8. Cumulative record showing the "scalloped" pattern of responding commonly observed in subjects responding under relatively long fixed-interval schedules of reinforcement. Such a pattern is marked by little or no responding early in the interval (i.e., after a reinforcer is delivered), followed by a gradual increase in responding until a high rate is reached at the end of the interval. From Dews (1955). Copright 1955 by the Society for Experimental Therapeutics. Reproduced by permission.

Characteristics of Good Graphs

Good graphs are accurate, clear, complete, and meaningful. As Tufte (1983, p. 53) asserts, "Graphic excellence begins with telling the truth about the data." An *accurate* graph tells the truth; it presents the actual findings of an experiment without distortion or misrepresentation. In the interest of accuracy, it is the researcher's responsibility to ensure that (1) the data are not misplotted, (2) the labels accurately describe the parts of the graph, and (3) the visual representations of the numbers are directly proportional to the numerical quantities represented (Tufte, 1983; see Sidebar 7–1).

Although accuracy is an absolute necessity, even a perfectly accurate graph is of minimal value as a judgment aid if it does not portray the data clearly. A *clear* graph is one that is easy to interpret. That is, it portrays findings in a way that allows viewers to discern easily the relation between the independent and dependent variables. In general, the clarity of a graph can be compromised in three ways. One is the presentation of too much information in a single figure. What constitutes "too much" information can be determined only empirically—one should always check the clarity

Sidebar 7-1
What Is Graphic Distortion?

As Tufte (1983) explains, a graph is distorted when the visual representation of the data is inconsistent with the numerical quantities represented. The figure below, which appeared in the *New York Times*, portrays the increase in fuel economy standards set by the U.S. Congress and the Department of Transportation from 1978 to 1985. (This graph might be visualized as a bar graph stood up on its right side, turned 90° clockwise, and laid back nearly horizontal, with some [but not all] expansion joints on the highway representing miles per gallon.)

Do you see any problem with this graph? In it, the change in numerical quantity (miles per gallon) is represented visually by changes in the length of certain lines (expansion joints) on the highway. To be free of distortion, the percentage change in each line length should be equal to the percentage change in miles per gallon. This is not the case. The difference between the length of the first line (1978) and that of the last line (1985) is 4.7 inches, representing a 783% increase. Hence, this difference should represent a 783% increase in miles per gallon from 1978 to 1985. To be consistent with the lines in the graph, the fuel economy standard would have had to be 158.94 miles per gallon in 1985. In fact, the actual standard set for 1985 was only 27.5 miles per gallon, a 53% increase. This graph is a clear (and extreme) example of graphic distortion. The graph grossly exaggerates the increase in fuel economy standards, leading the viewer to conclude erroneously that the U.S. Congress and Department of Transportation were raising their standards at a fantastic rate.

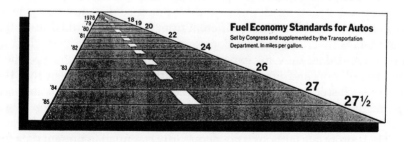

Source: *The New York Times*, August 9, 1978, p. D2.

of figures by presenting them to sympathetic colleagues—but it is usually difficult to respond easily to figures with more than four data paths.

A second detriment to clarity is unnecessary decoration (see Figure 7.9). Features of a graph that do not directly aid in interpretation are useless clutter. They make the viewer's task more difficult and should be avoided.

Constructing a graph in a way that distorts actual changes in behavior also reduces clarity and may lead the viewer to erroneous interpretations. In particular, the apparent magnitude of the effect of an intervention can be dramatically influenced by manipulating the scaling of the vertical and horizontal axes (i.e., the aspect ratio). For example, treatment effects can be exaggerated by restricting the range of the vertical axis or minimized by lengthening the range. In Figure 7.10, the apparent difference between performance in baseline and treatment conditions is much more impressive in the top graph than in the bottom graph, although the same data are portrayed in each.

A *complete* graph is one that presents all the data collected on the particular behavior represented in the graph. If, for example, 40 measures of a target behavior were collected during the course of an experiment, a complete graph of those data would include all 40 measures, not selected portions of those measures. Graphs should not present data out of the context of the totality of observations if doing so leads the viewer to different conclusions than if all the data were presented. For example, what appears to be a clear intervention effect when selected data are presented as in the top graph of Figure 7.11 becomes less clear when a complete data set is displayed as in the bottom graph. It would therefore not be legitimate to present the data displayed in the top graph as a fair representation of the complete data set.

As a rule, it is legitimate to present portions of data sets provided (1) there is good reason for doing so and (2) doing so does not lead one to draw conclusions dramatically different from those one would draw if the entire data set were presented. For example, many studies published in JEAB present data only for experimental sessions that meet a specified stability criterion and ignore data from earlier sessions. Given that the focus of such studies is on steady-state (i.e., stable) behavior, this strategy is perfectly satisfactory, although researchers should always check to see whether there are interesting patterns of behavior prior to stability. If there are, it is appropriate (although not necessarily essential) to present data for all sessions.

Now and again in the course of experiments, one obtains data that differ substantially from other observations made under similar conditions. All behavioral researchers must face the problem of dealing with

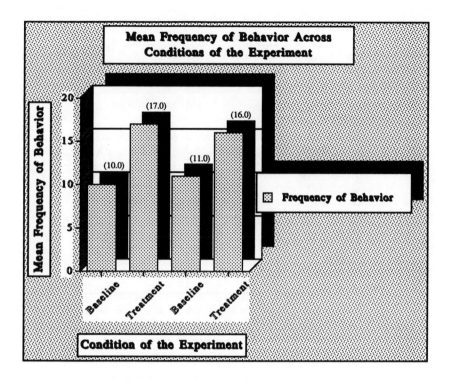

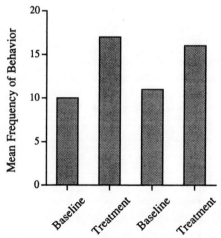

FIGURE 7.9. Two bar graphs of the same hypothetical data set. The top graph exemplifies the use of unnecessary and distracting decoration. For instance, the numbers printed above each bar, the title at the top of the graph, the legend, and the horizontal axis label are redundant. Moreover, the three-dimensional depth, various shadings, and graph frame are a waste of ink. The bottom graph provides a much less cluttered and thus more effective portrayal of the data.

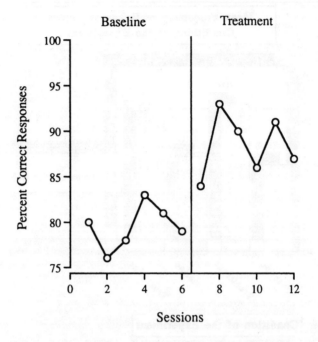

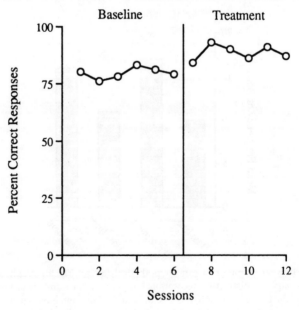

such "outlying" or "anomalous" observations. Completeness in reporting dictates, of course, that such data be reported. This is the srecommended strategy *unless there is an obvious reason for the anomaly*. If there is a clearly specifiable reason for the anomaly, it is legitimate to discard those observations, for they are interpretively worthless. As Johnston and Pennypacker (1993b, p. 307) relate:

> The reasons why some data might be interpretively worthless are familiar to any investigator. A subject may have been ill or emotionally upset. Known extraneous influences may have occurred in spite of the investigator's best efforts. Measurement procedures may have been unacceptable for some reason. Equipment may have malfunctioned. The factors constituting the control or experimental condition may have varied beyond acceptable limits. The list sometimes seems endless. What defines data as troublesome ["anomalous" in our terms] is that they were not collected under the same conditions as were the data that do qualify for analysis. What qualified data for being discarded is that they were either collected in the presence of undesirable variables or that some aspect of their collection was unacceptable.

Although it is legitimate to discard data when there is a good reason to do so, this strategy is acceptable only if it is used very infrequently. Sidman (1960, p. 187) makes this point convincingly:

> If uncontrolled variability occurs only rarely, [the experimenter] may justifiably not even mention the data in question. No colleague is as informed about the possible sources of occasional uncontrolled variability in a given laboratory as is the experimenter who works in that laboratory. He [or she] is in the best position to evaluate such instances, and he [or she] cannot pass the responsibility on to others. If the variability occurs frequently enough to be a serious problem, none of the data should be reported until the sources of the variant data have been eliminated. There is no middle ground.

In a sense, a graph can be considered incomplete if it depicts summary measures (e.g., means of the performance of several subjects, means of the observations within phases), especially if it does not include corresponding measures of variability (e.g., standard deviations, ranges). A graph is also incomplete if it is not fully labeled. Graphs should contain as many labels as necessary to facilitate the viewer's understanding of the relation that the graph represents. As mentioned above, however, superfluous labels that merely decorate a graph only reduce clarity.

Finally, graphs should be *meaningful*. The meaningfulness of a graph

FIGURE 7.10. Two graphs of the same hypothetical data set showing the effect of restricting the range of the vertical axis on the apparent size of the intervention effect. The range of the vertical axis in the top graph is restricted to 25% (75–100) of the total possible range (0–100), and the apparent size of the intervention effect is much greater than that portrayed in the bottom graph, in which the range of the vertical axis is not restricted.

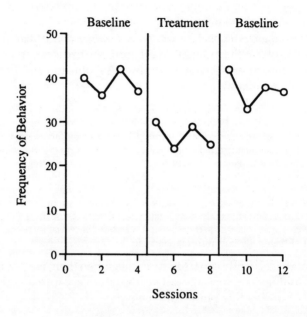

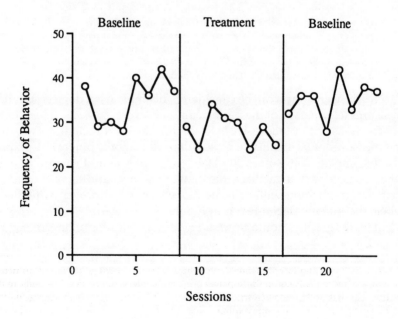

refers to the relevance and adequacy of the answer it provides to the experimental question, or of other useful information about the behavior of interest and the variables that influence that behavior. The dependent variables of interest in an experiment are dictated by the experimental question, and these variables should be the primary focus of analysis. But the researcher need not focus solely on these variables, for such a narrow focus may preclude the disclosure of other important relations (Johnston & Pennypacker, 1993b). For example, behavioral pharmacologists are often interested in the effects of drugs in nonhumans (e.g., rats, pigeons, monkeys) responding under a simple schedule of reinforcement, such as a fixed-ratio (FR) schedule. An experimental question such as "What are the effects of cocaine on the responding of rats under an FR 25 schedule of food delivery?" dictates that they measure overall response rate as a dependent variable and focus on that measure when they analyze their findings.

Other variables related to overall rate of responding would also be important, such as postreinforcement-pause time and running response rate (rates calculated with postreinforcement-pause time excluded). Analysis of the drug's effects on these variables allows one to discern how the drug affected overall rates (e.g., by increasing pausing or decreasing running rate). In this example, overall rate of responding is the primary dependent variable dictated by the experimental question. Postreinforcement-pause time and running rate are secondary dependent variables that the researcher measures in order to elucidate changes in the primary dependent variable, thus providing a more comprehensive analysis of the drug's effects.

With automated data-recording devices, it is tempting to measure a wide variety of dependent variables in the hope that one or more of the variables measured will change as an orderly function of changes in the independent variable. This strategy has little to recommend it. As a rule, *in basic research one should measure only the variables that are dictated by the experimental question and any other variables that may account for changes in the primary dependent variables. In applied research, it also is judicious to measure outcomes that provide an index of possible adverse side effects of treatment.*

←——————————————————————————————————————

FIGURE 7.11. Two graphs of the same hypothetical data set. The top graph depicts only portions of the entire data set (the last four observations in each phase) and consequently shows clear shifts in performance with no overlapping variability across conditions (i.e., no values in the treatment phase lie within the range of baseline values). When the entire data set is plotted, as in the bottom graph, the shifts in performance across phases are less clear, due to the presence of overlapping variability across conditions.

Constructing Graphs with Computers

The care with which graphs should be prepared depends, in part, on the purpose for which they are intended. Constant evaluation of incoming data is characteristic of much behavior analytic research, and graphs play an invaluable role in this process. Often, a researcher prepares many different kinds of graphs over the course of an investigation. These graphs are intended only to influence members of the research team and will not be shared with a larger scientific community. Therefore, they can be relatively crude, and need not follow all the conventions described previously.

Some 15 years ago, when labor-intensive drafting techniques were used to construct high-quality graphs, everyday working graphs often were little more than a hand-lettered sheet of graph paper with data points and data paths. Although such graphs have generated a wealth of representations of behavior analytic data and continue to do so, a graph's value as a judgment aid is improved if the graph is clear, accurate, and easy to read. Fortunately, excellent graphs can now be prepared quickly and easily through the use of a personal computer using appropriate graphics software. For example, the figures in this chapter were created easily on a Macintosh computer using Cricket Graph III (by Computer Associates International). Various types of relatively inexpensive graphics software are available. The manuals for such software usually are written at a rudimentary level, and most people find it easy to learn to construct high-quality graphics. *Computer proficiency is a necessary skill for behavioral researchers, and students are well advised to develop it early in their careers.*

Interpreting Graphs

As Cooper et al. (1987) note, before evaluating the data displayed in a graph prepared by someone else, one should carefully read all the labels on the graph as well as the figure caption to be clear about what the graph is displaying. One should also examine the graph's construction, noting the range and scaling of the axes. Once one understands what the graph portrays and how it is portrayed, one can begin to interpret the patterns of performance displayed in the graph.

When one interprets a graph, one makes a decision as to whether a functional relation exists between the independent and dependent variables. The decision one reaches is controlled by various features of the data display. Therefore, one can view a graph as a complex stimulus that controls the analytical verbal behavior of the researcher (Parsonson & Baer, 1986). This verbal behavior involves statements such as "Yes, the intervention was effective," "No, the intervention was not effective," or "I'm not sure whether or not the intervention was effective." The stimulus properties of graphs that should control a researcher's statements about the data include means, trends, variability, and latency of change.

A *mean* is the average level of performance across a specified number of observations. For example, when one considers mean levels of performance to assess the effects of an applied intervention evaluated using an A–B–A–B design, one compares the average level of performance across the baseline (A) and intervention (B) phases. To make this comparison, one can simply estimate visually the average level of the dependent variable across each phase and determine the degree to which the level shifted as the phases changed. In general, the larger the shift in mean performance across phases, the clearer the intervention effect. A more precise way of visually examining changes in means across phases is to calculate the mean for the observations in each phase and draw a solid horizontal line on the graph representing the mean (see Figure 7.12). This technique is

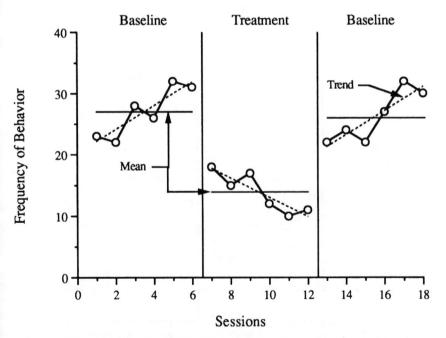

FIGURE 7.12. A graph of hypothetical data showing changes in means and trends across phases of a withdrawal design. Solid horizontal lines drawn through the data points in a phase depict the mean level of performance during that phase. Dashed diagonal lines drawn through the data points of a phase depict the trend of performance during that phase.

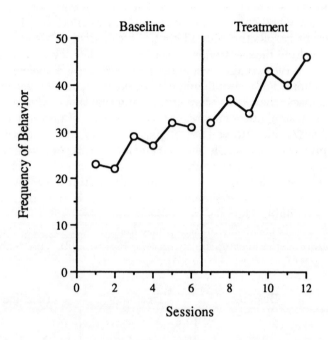

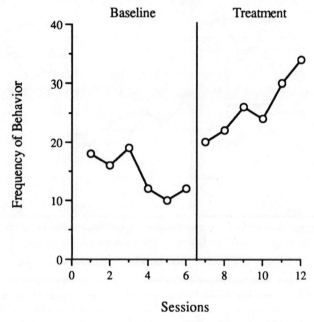

especially useful when data are highly variable within and between phases. Figure 7.12 illustrates hypothetical data showing changes in means across phases of a withdrawal design.

A *trend* is a systematic increase or decrease in data over time. Trends are considered when it appears that the direction of behavior change is altered as the intervention is applied and withdrawn (Kazdin, 1982). To consider trends in assessing intervention effects, one visually estimates the direction and degree of slope in the data across adjacent phases of the experiment drawing dashed trend lines on the graphs (see Figure 7.12) may make this task easier. In general, an intervention effect is clear if the intervention produces a change from no trend during baseline to a trend (increasing or decreasing) during the intervention phase, or if a trend reverses in direction when conditions change. The hypothetical data in Figure 7.12 illustrate changes in trends (in addition to changes in means) across phases.

Trends during baseline conditions may or may not pose interpretive difficulty. Figure 7.13 presents graphs of two hypothetical data sets showing the effects of an intervention designed to increase the level of behavior. Trends are evident in the baseline phases of both data sets. The baseline trend in the top graph is problematic because the intervention was intended to change behavior in the same direction as the trend during the baseline phase. In this case, it is possible that behavior would have continued in that direction even if there had been no intervention. The baseline trend in the bottom graph is not problematic because the intervention was intended to change behavior in a direction opposite that of the trend during the baseline phase.

Variability is the extent to which the measures of the dependent variable differ from one another. In considering variability, one determines how much the data points in a graph are spread out along the vertical axis. The extent to which data overlap across phases is of primary concern in evaluating variability. For example, Figure 7.14 shows two data sets obtained from a withdrawal design. In the top graph, the variability within the baseline phases overlaps the variability within the treatment phases (as indicated by the shaded area). Despite the changes in means, the interven-

←——

FIGURE 7.13. Two graphs depicting hypothetical data showing baseline trends in performance. In both cases, treatment was intended to increase the frequency of behavior. In the top graph, the increasing trend during baseline poses interpretive difficulty because one could argue that the frequency of behavior would have continued to increase without intervention. In the bottom graph, the decreasing trend during baseline poses no interpretive difficulty, since one would predict behavior to continue to decrease in the absence of treatment. The reversal of this baseline trend by the treatment provides strong evidence of the treatment's efficacy.

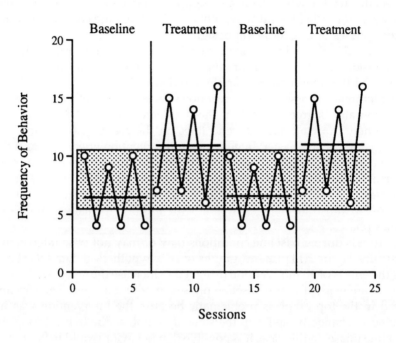

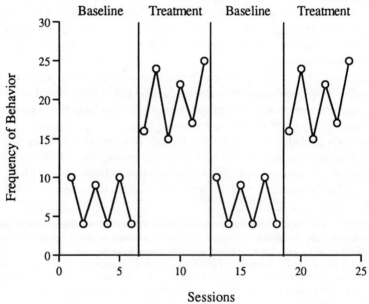

tion effect in the top graph is not nearly as clear as the intervention effect in the bottom graph, in which no overlapping variability across phases is evident.

When a graph presents summaries of raw data, it is crucial to consider whether any measure of variability is presented. If not, as would be the case if only means were presented, it is very difficult to make an informed decision concerning the effects of the intervention. Because it is difficult to respond meaningfully to summary measures of behavior (e.g., means, modes, medians) without corresponding measures of variability, it is convention to present some measure of variability (e.g., range, standard deviation, standard error of the mean) whenever summary data are shown.

Latency of change refers to the time between a change in conditions (e.g., applying or withdrawing the intervention) and a change in behavior. In general, interpretation of an intervention effect is made much easier if a change in behavior occurs immediately following a phase change. The longer the latency to the observation of a change in behavior, the less clear the intervention effect.

As noted by Cooper et al. (1987), delayed changes in performance can indicate one of two things. First, delayed effects may indicate that the intervention (e.g., an antidepressant drug) must be in effect for a period of time before it begins to affect behavior. Second, a long latency may indicate that an extraneous variable became operative at the point at which behavior began to change. For example, the latent change in behavior evident during the intervention phase in the top graph of Figure 7.15 is not as clearly attributable to the intervention as is the change in behavior evident in the bottom graph, given that it is not necessary that the intervention be in place for some period of time before it can influence behavior.

Tables

Data can be visually displayed in tables as well as in graphs. *Tables* usually comprise lists of numbers that represent values of the dependent

←———

FIGURE 7.14.   Two graphs portraying hypothetical data obtained from a withdrawal design. In the top graph, horizontal lines through the data points indicate mean levels of performance. The shaded area indicates overlapping variability across phases (i.e., the range of values of the target behavior that are observed across all the phases). Changes in means notwithstanding, the treatment effect is not clear in the top graph because it could not be replicated reliably; during half of the treatment sessions, an increase in the frequency of behavior beyond the range of baseline values was not observed. In the bottom graph, none of the values obtained during treatment lies within the range of baseline values (i.e., no overlapping variability is present). Thus, the treatment effect is relatively unambiguous.

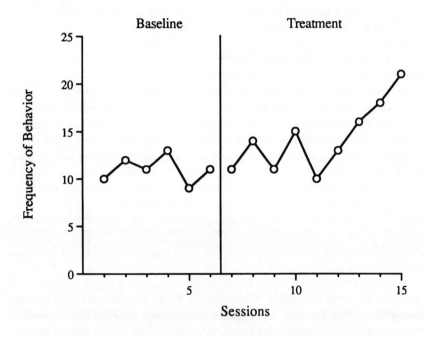

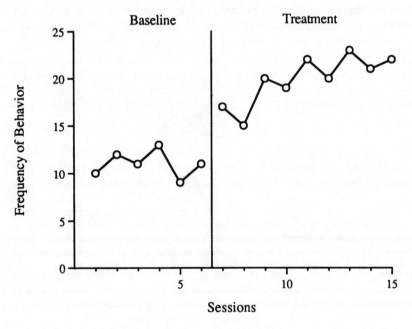

variable across conditions of the experiment. Table 7.2 is a hypothetical example of a table used to present data from an experiment employing a withdrawal design. The primary limitation of tables is that they cannot effectively convey patterns in the data. In cases in which it is important that interpretations be controlled by patterns in the subjects' performance (e.g., trends, variability), graphs are far superior to tables. Moreover, large tables with hundreds of values are intimidating and require several minutes of analysis to discern relationships between the variables of interest. Smaller tables are easier to interpret, but it is often possible to report their contents in the text, making them redundant. In behavior analysis, tables are primarily used as archival repositories of raw data, as in the Appendices provided in JEAB articles, not for presenting primary data.

Summary and Conclusions

Results of a study are *experimentally significant* to the extent to which there is a reliable relationship between an independent and a dependent variable. They are *therapeutically* (or *clinically*) *significant* to the extent to which the changes in behavior produced by an independent variable are socially significant. Social validation procedures (discussed in Chapter 3) are used to evaluate therapeutic significance, which is important in applied research, but not in basic, research (Kazdin, 1982; Risley, 1970).

Experimental significance is assessed via statistical or graphic analysis of the results of a study. Although it may seem unacceptably loose and imprecise to scientists without appropriate histories, graphic analysis is common in science (Tufte, 1983) and has characterized behavior analysis since the inception of the field. Several authors have discussed at length the advantages of graphic analysis for a science of behavior (e.g., Cooper et al., 1987; Johnston & Pennypacker, 1993b; Parsonson & Baer, 1986; Sidman, 1960). These advantages are real and have led to forceful arguments in favor of the method. For example, in concluding an excellent chapter dealing with graphic analysis, Parsonson and Baer (1986, p. 184) offered this forthright opinion:

←——————————————————————————————————————

FIGURE 7.15.   Two graphs of hypothetical data obtained under baseline and treatment conditions. The top graph depicts a latent increase in performance during the treatment condition (several sessions of treatment were required before increases in performance relative to baseline were observed), while the bottom graph depicts an immediate increase in performance upon introduction of treatment. In general, latent treatment effects are more difficult to interpret than are immediate ones.

TABLE 7.2. Hypothetical Data Showing the Effects
of an Intervention on Mean Duration of a Target
Behavior

Subject	Mean duration (min)			
	Baseline	Treatment	Baseline	Treatment
1	10	176	22	182
2	12	124	11	128
3	0	132	0	156
4	5	110	10	105
5	15	164	13	158
MEANS:	8.4	141.2	11.2	145.8

We believe that the fine-grained analysis of data ... provides researchers and research audiences with information and analytic powers that cannot be matched by statistical analysis, or, indeed, in any other way. ... We believe that the graphic analysis of an ongoing graphic data presentation interacts power-fully with single-subject ["within-subject" in our terms] design to produce a responsive, functional, accurate, and analytic approach to the scientific investigation of behavior, just as argued so long and so effectively by Skinner (e.g., 1956) and by Sidman (1960).

Many behavior analysts, including the authors of this volume, agree with Parsonson and Baer. Constructing good graphs and responding to them in ways that are supported by the appropriate community of scientists are acquired skills. Students of behavior analysis should develop them early. One reasonable strategy to help students learn the conventions of graphic data analysis is matching-to-sample; that is, students should model their graphs after those typically used in their field of specialty and their interpretations of those graphs after the conclusions that appear in published articles. Hands-on training by established behavioral researchers will also prove invaluable.

CHAPTER 8

Statistical Analysis of Data

A *statistic* is a number that describes some characteristic of a set of data. Descriptive statistics, such as means and standard deviations, summarize sets of data collected under the same experimental condition. Inferential statistics, such as *t*-tests and *F*-ratios, summarize differences in sets of data collected under different experimental conditions. A wide range of descriptive and inferential statistics are available, and no attempt will be made to discuss them here. Instead, this chapter considers the general functions of descriptive and inferential statistics and offers eight practical suggestions for the appropriate use of statistics in behavior analytic research.

Measures of Central Tendency and Measures of Variability

Descriptive statistics include measures of central tendency and measures of variability. Measures of central tendency deal with the point on a scale at which a data array is centered. Three commonly used measures of central tendency are the mean, the median, and the mode. The *mean* is the average of the scores in a distribution; it is calculated by summing the individual scores and dividing by the number of scores. The *median* is the midpoint value in a distribution. Half the scores fall above this point, half below. The *mode* is the most commonly occurring score in a distribution. When a distribution of numbers is completely symmetrical, the mean and the median will be the same score. Further, when the distribution is unimodal, the mode will be the same value as the other two measures.

Measures of variability describe the amount of deviation around a measure of central tendency (usually the mean) of a distribution. Three commonly used measures of variability are the range, the variance, and the standard deviation. The *range* is computed by subtracting the lowest from the highest observed score. The *sample variance* is the sum of the squared deviation scores (a deviation score is the difference between an individual observation and the mean for the group of scores from which that observation was taken). The *standard deviation* is the square root of the sample variance.

As a rule, the use of measures of central tendency without accompanying measures of variability is likely to be misleading and should be avoided. Determining the measures of central tendency and variability that are appropriate for describing a given set of data requires consideration of the characteristics of the data distribution, which are usually apparent upon visual inspection of a plot of the raw data. It also requires knowledge of the characteristics of the statistics under consideration, which can be obtained from any good statistics book (e.g., Bordens & Abbott, 1991; Howell, 1992).

Characteristics of Samples and Populations

In many experiments, a researcher is interested in a relatively large group of scores, termed the *population*, but is able to collect only a subset of those scores, termed the *sample*. In between-subjects designs, the population typically refers to scores from a large group of individual subjects that share some defining characteristic. In within-subject designs, the population characteristically refers to an infinitely large number of scores that particular subjects (those studied) could generate under a given experimental condition.

Inferential statistics allow researchers to arrive at some probability statement concerning the state of affairs in a population given the sample data collected. In most cases, they determine the probability of obtaining differences in sample values (e.g., differences between the means of two or more groups) of at least the amount observed if there is no true difference between the values of the populations from which those samples were drawn. This information goes beyond that provided by descriptive statistics. As Howell (1992, p. 59) explains, "Descriptive statistics will not tell us, for example, whether the difference … between two obtained sample means is small enough to be explained on the basis of chance alone or whether it represents a true difference that might be attributable to the effect of our experimental treatments." Inferential procedures are employed to determine whether the observed differences are large enough to be the result of something other than chance.

Inferential statistics estimate the value of some characteristic (termed a *parameter*) of a population given the value obtained in a sample. Two basic options are available for estimating population values: point estimation and interval estimation. The former is more common, although many contemporary statisticians recommend the latter. According to statistical theory, a point estimate, such as the mean of a sample distribution, is an unbiased estimator of the true but unknown population value. That is, it is the single best guess of the value (e.g., the mean) for the whole population to which we do not have direct access.

Because we are testing only a subset of the population, however, the value derived from the sample data will not correspond exactly to that of the whole population. Each time we draw a new sample and gather data, we will produce a different value for the summary measure. The overall average of the summary measures gathered from a large number of samples will be approximately equal to the population value. The difference between a statistic and its corresponding value in the population parameter—e.g., for example, the difference between the population mean and the sample mean—is called *sampling error*. It is sampling error to which we refer when we say that a difference in sample means is due to chance.

In addition to point estimation, interval estimation is another approach to making inferences about true but unknown population values. A *confidence interval* traps the population value, with some level of certainty, between an upper and a lower limit. In the point estimation approach, we derive a single number with which to estimate the state of the population. In the interval estimation approach, we allow for some variation and construct a range of possibilities for the population value. In constructing confidence intervals, the researcher is recognizing sampling error and is using it to produce a range of values to describe the probable state of affairs in the population.

Confidence intervals tell readers something about the amount of sampling error in a study. For example, a confidence interval on a mean is constructed with a lower limit of 2.4 and an upper limit of 9.7. The original scale used to collect that data was 0 to 10. In this case, the range of the confidence interval is almost as large as the original scale of measurement. This tells us that there is a lot of sampling error involved and that tight experimental control was not achieved. On the other hand, if the original scale ranged from 0 to 100 and the confidence interval is 22.4 to 29.7, the picture is different. In this case, the confidence interval on the mean is narrow compared to the original scale of measure. This interval suggests that tight experimental control was achieved and that sampling error was at a minimum.

True state in the population

Decision based on sample data:	No difference (Null hypothesis is true.)	Difference (Null hypothesis is false.)
There is a true difference (reject the null hypothesis).	Type I error $p = \alpha$ (alpha) **1**	Correct decision **2**
There is not a true difference (retain the null hypothesis).	Correct decision **3**	Type II error $p = \beta$ (beta) **4**

FIGURE 8.1.   Possible decisions based on inferential statistics.

Type I and Type II Errors

Comparisons of sample means via inferential statistics allow a researcher to make tentative decisions about population means. Two decisions are possible: There is a difference or there is not a difference. The proposition that there is no difference in population means is termed the *null hypothesis,** and the researcher can either accept or reject it. If the researcher rejects the null hypothesis when it is actually true, a *Type 1* error has been committed. If the researcher accepts the null hypothesis when it is actually false, a *Type 2* error has been committed. Figure 8.1 depicts Type 1 and Type 2 errors, as well as the two kinds of correct decisions about population parameters that can be made on the basis of sample data.

Type 1 and Type 2 errors are illustrated by quadrants 1 and 4, respectively, in Figure 8.1. When inferential statistics are used, the probability of making a Type 1 error (termed α, Greek "alpha") is equivalent to the probability level (p) that one uses as a cutoff for accepting or rejecting the null hypothesis. In most studies, an α level of 0.05 or 0.01 is used, meaning that there is a 5% chance or 1% chance of a Type I error. The probability of making a Type 2 error is termed β (Greek "beta"). The probability of making a Type 2 error increases as the probability of making a Type 1 error

---

*Many other forms of the null hypothesis are possible. For example, measures other than means may be of interest, and the proposed difference in population values may be something other than 0. As an illustration, one might postulate as a null hypothesis that the modal value for alcohol consumption in college freshmen is 1.2 times as great as the modal value for alcohol consumption in college professors. For simplicity, we will consider the null hypothesis only in terms of mean comparisons with no difference expected.

decreases, but β is not something that is set by the researcher when conducting a statistical test. In fact, it typically remains an unknown quantity and is simply ignored (Howell, 1992).

Quadrants 2 and 3 in Figure 8.1 represent correct decisions. The probability that the researcher correctly rejects the null hypothesis, shown in quadrant 3, defines the *power* of a statistical test, and it is inversely related to β. Power is affected by the chosen α level, sample size, and whether a one- or two-tailed test is used (a one-tailed test rejects the null hypothesis only if the difference in means is in a specified direction, e.g., if the mean for the control group exceeds the mean for the experimental group). Power increases as the α level is reduced, and two-tailed tests are less powerful than one-tailed versions of the same tests. The power of all inferential statistics increases as the number of observations in the samples increases, and procedures are available for determining the number of observations needed for a particular test to have specified power given a particular anticipated effect size (Gravetter & Wallnau, 1990; Keppel, 1982). There is no agreed-upon standard for the proper amount of power (Keppel, 1982). Low power in statistical analysis, however, is a common problem in psychology in general (see Cohen, 1977; Rosnow & Rosenthal, 1989b; Sedlmeier & Gigerenzer, 1989), and behavior analysts who intend to use inferential statistics should consider the power of their proposed tests.

The terms Type 1 error and Type 2 error are sometimes taken to mean more than inappropriately rejecting or accepting the null hypothesis. Rejection of the null hypothesis suggests that an independent variable affected a dependent variable; therefore, Baer (1977, p. 170) indicates, "[t]o make a Type 1 error is to affirm that a certain variable is a functional one, when in fact it is not." Following the same logic, he indicates (p. 170) that "[t]o make a Type 2 error is to deny that a certain variable is a functional one, when in fact it is." Kazdin (1982, p. 242) defines Type 1 and Type 2 errors in a similar fashion:

> Type 1 error refers to concluding that the intervention (or variable) produced a veridical [actual] effect when, in fact, the results are attributed to chance. Type 2 error refers to concluding that the intervention did not produce a veridical effect when, in fact, it did.

Such definitions are common. But one must understand that they encompass more than data analysis errors. As a hypothetical example, assume that a researcher uses an A–B design to evaluate the effects of a hypertensive person's wearing a copper bracelet on that person's blood pressure. The subject does not wear the bracelet for the first 60 days of the study (A), but does wear it for the final 60 days (B). The participant's blood pressure is recorded each day. It falls from an average of 187/123 mm Hg during the A phase to 119/87 mm Hg during the B phase. A paired t-test reveals that this effect is significant at an α level of less than 0.001, and the effect is also readily evident upon visual inspection of the graphed data.

Given this finding, the researcher rejects the null hypothesis and concludes that wearing of the copper bracelet was effective in treating hypertension. In fact, unknown to the researcher, the participant sought medical treatment for the high blood pressure just before the B phase began, and it was this variable, not the bracelet, that was responsible for the fall in systolic and diastolic pressure. By erroneously attributing the effect to the intervention (bracelet), the researcher made a Type 1 error in the sense that the term was used by Baer (1977) and Kazdin (1982). That error is not due, however, to incorrect rejection of the null hypothesis. Errors of inference can occur at all stages of data evaluation, not just in determining whether or not dependent measures actually differ across conditions.

The probability of committing a Type 1 error is specified when inferential statistics are used to analyze data, but not when the analysis involves visual inspection of graphed data. As noted in Chapter 7, Baer (1977) and others have argued that graphic analysis as performed by behavior analysts minimizes Type 1 errors and that this strategy is appropriate for a science of behavior concerned with isolating independent variables (treatments) that produce large and consistent behavioral effects. A focus on such variables is certainly advantageous in applied behavior analysis. Variables that produce small but reliable effects, however, may also be important.

There are no data to indicate whether statistical data analysis procedures are better or worse than graphic procedures for detecting small, but meaningful, treatment effects (Huitema, 1986b). It is widely recognized that inferential statistics often identify small effects, and indeed they do. In fact, if the samples are large enough, trifling differences in means will be significant at p values less than 0.05 (e.g., Kupfersmid, 1988; Lykken, 1970). Cohen (1990) provides a superb example of how statistically significant findings can be of no practical or theoretical importance. The example is based on an actual study in which the heights and IQ scores of 14,000 children were compared via correlational methods. The resultant correlation coeffecient was 0.11, which is significant at an α level of less than 0.01. The theoretical significance of this relationship is nonexistent, and its practical value is no greater. As Cohen (1990, p. 1309) pointed out, "For a 30-point increase in IQ, it would only take enough growth hormone to produce a 3.5-ft increase in height, or with the causality reversed, a 4-in increase in height would require an increase of only 233 IQ points."

No one should use statistics to search for meaningless relationships between variables. But even researchers who know better may be inclined to do so when no strong and meaningful relations are apparent in the results of a completed study. The temptation is likely to be especially strong when that study will be considered in evaluating the researcher for a graduate degree, an academic promotion, or a grant award. Avoid the temptation. Statistics can in some cases clarify and amplify the findings of a study, but they cannot impose order on chaos or create meaningful functional relations where none exists.

Statistics as Judgment Aids

Like graphs, statistics are judgment aids (Michael, 1974). Statistics simplify and abbreviate experimental results, which makes it easier for scientists and other people to react to those results, and this is the primary advantage of statistics. As abbreviations, however, statistics lack some stimulus aspects of the raw data, and if one's entire reaction is based on the abbreviations, the missing features cannot affect behavior. Therefore, conclusions based on the statistic alone may be inaccurate or incomplete.

As an example of how statistics abbreviate findings, consider the results of a hypothetical two-groups experiment examining the effects of a moderate dose of alcohol on the reaction time of college students. Before the study was conducted, 40 students were selected at random from all psychology majors at East Goose Community College. Of these 40 students, 20 were randomly assigned to the experimental group and drank 2 ounces of 90 proof vodka in orange juice 30 min before being tested. The remaining 20 students were assigned to the control group; they drank pure orange juice 30 min before their reaction time was measured. The same test of reaction time was used with all students, and each student was tested once. Raw data from the experiment, in the form of the reaction time recorded for each subject, are presented in Table 8.1.

What can you conclude from a visual inspection of the raw data in Table 8.1? Not much, without difficult and protracted analysis. Any 40 numbers presented in tabular form are difficult to react to in any meaningful way unless there are large and consistent differences across conditions. Figure 8.2, which depicts the mean reaction time (and standard deviation) for each group, greatly abbreviates the raw data in Table 8.1 and makes it much easier to react to the results of the study. Clearly, the mean reaction times for the two groups are different; on average, students who drank vodka reacted 0.6 sec slower than students who drank orange juice.

But can we be sure that the difference is due to the vodka? No. Even though the subjects were randomly selected and assigned to groups, we

TABLE 8.1. Reaction Times of Hypothetical Subjects
Who Consumed Vodka in Orange Juice
(Experimental Group) or Orange Juice Only
(Control Group)[a]

Reaction times (sec)									
Experimental group					Control group				
1.3	2.4	3.1	1.9	4.6	1.2	1.1	0.9	2.3	2.6
2.4	3.3	1.1	1.8	3.2	1.7	2.4	1.8	1.4	3.7
0.9	3.1	2.6	1.9	2.3	1.9	2.0	0.7	2.4	3.0
4.1	2.6	1.8	3.2	3.5	3.4	1.3	1.7	2.2	1.8

[a]At 30 min before being tested, students in the experimental group
drank 2 ounces of 90 proof vodka in orange juice; students in the
control group drank orange juice only.

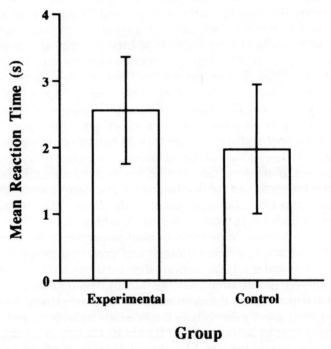

FIGURE 8.2.   Mean reaction times and standard deviations for hypothetical groups of college
students who did (experimental group) and did not (control group) consume alcohol before
testing. Each group comprised 20 students; the individual reaction times for the students are
presented in Table 8.1.

would expect some difference in the sample means had all subjects been treated identically. By calculating an appropriate inferential statistic, such as a t-test, we can evaluate the probability of obtaining a difference in sample means of 0.6 sec or more, given that there was no difference in population means. The proposition that there is 0 difference in population means is the null hypothesis, and it is this hypothesis that is tested when our sample means are compared via a t-test. For the data in Table 8.1, the probability of obtaining a mean experimental reaction time that exceeded the control reaction time by 0.6 sec or more was less than 0.01. This probability (p) is based on an obtained one-tailed t-test of 2.06 with 38 degrees of freedom.

What does this p value indicate? Only the probability of obtaining the results that were obtained in the study, given that the null hypothesis actually is true. There is long-established precedent to accept as statistically significant p values of less than 0.05 (Rosnow & Rosenthal, 1989b). If we follow this precedent, the obtained difference in reaction times is significant. That is, the null hypothesis (in this case, that there is 0 difference in population means) can be rejected, and there is reason to believe that the difference in means is due to the different conditions to which the two groups were exposed.

For many researchers, statistically significant results are important. As Rosnow and Rosenthal (1989b, p. 1277) humorously put it:

> It may not be an exaggeration to say that for many PhD students, for whom the .05 alpha has acquired almost an ontological mystique, it can mean joy, a doctoral degree, and a tenure-track position at a major university if their dissertation p is less than .05. However, if the p is greater than .05, it can mean ruin, despair, and their advisor's suddenly thinking of a new control condition that should be run.

Rosnow and Rosenthal go on to explain that there is nothing sacred about a p value less than 0.05, or any other level, and that dichotomous significance testing has no logical basis. Moreover, as emphasized by many authors (e.g., Cohen, 1990, 1994; Morrison & Henkel, 1970; Tukey, 1991), null hypothesis testing is widely misunderstood and is not particularly useful.

Some of the most common misunderstandings can be readily illustrated by example. Consider two hypothetical experiments, each intended to examine the effects of systematic desensitization in treating a spider phobia exhibited by college students. Each experiment used a randomized two-group design with 25 students in each group. The systematic desensitization procedures used in the two studies were similar, but differed in small ways. In both experiments, the mean distance from the spider tolerated by students in the experimental group was less than the mean distance tolerated by students in the control group. When the data were analyzed via *t*-test, the obtained *p* was exactly 0.05 in Experiment A and 0.01 in Experiment B. Given these findings, which of the following conclusions would you declare to be true?

1. The systematic desensitization procedure used in Experiment A produced a larger effect than the procedure used in Experiment B.
2. The systematic desensitization procedure used in Experiment A produced a more significant effect than the procedure used in Experiment B.
3. The probability of replicating the results of Experiment A is 95%, and the probability of replicating the results of Experiment B is 99%.
4. The probability that the null hypothesis (that there is 0 difference in population means) is true is 0.05 in Experiment A and 0.01 in Experiment B.
5. The systematic desensitization procedures used both in Experiment A and in Experiment B produced clinically beneficial effects.

The correct answer is "None of them." All that the *p* values tell us in this example is the probability of obtaining a difference of at least the magnitude of that observed between sample means based on 25 observations each if the difference in population means is 0. This information is of little use in determining whether the results of a study are replicable, or even important. The replicability of results can be determined only by conducting further studies, and the importance of findings can be determined only through informed judgment based on much more than *p* values alone. Among the factors that influence the importance that scientists and laypeople assign to the results of a particular study are the nature of the problem under investigation, the procedures used to investigate that problem, the size and consistency of the effects obtained, and the relation of the obtained effects to those reported in similar investigations.

Suggestions for the Gainful Use of Statistics

Behavior analysts historically have not required ritualistic testing of the null hypothesis as a part of the research process. Statistics can be used, however, in other, more useful, ways. In some research areas, behavior analysts rely heavily on both descriptive and inferential statistics. For example, researchers dealing with the matching equation (described in Chapter 2) almost always report data that are averaged across sessions, then expressed as ratios. These ratios, which frequently are converted to logarithms, are then subjected to inferential testing using a regression analysis. This kind of thoroughgoing quantitative analysis is uncommon in other areas of behavior analysis, and skepticism concerning the general value of statistics in a science of behavior has been expressed by several authors (e.g., Baer, 1977; Michael, 1974; Sidman, 1960; Johnston & Penny-packer, 1993a,b). Other authors have questioned the appropriateness of particular inferential tests. For example, some have debated at length, without resolution, how, or even whether, time-series analyses should be used to analyze single-subject data (see Huitema, 1986a, 1988).

In general, beginning researchers should approach the use of statistics with caution. It is easy to err in selecting and interpreting statistical tests or to use them in situations in which they are unnecessary. In an effort to encourage neophyte researchers to use statistics conservatively, we offer eight general suggestions. In keeping with a conservative approach, we phrase each suggestion in terms of what the researcher should not do.

1. *Do not use descriptive or inferential statistics unless there is a good reason to do so.* One justifiable use of statistics is to summarize raw data that are otherwise too extensive or too complicated to react to in a meaningful way. Descriptive statistics are particularly useful for this purpose.

A second justifiable use of statistics is to satisfy the requirements of research consumers. Unfortunately, many people who review manuscripts submitted to journals and proposals submitted to funding agencies believe that only statistically significant results are important. As discussed in Chapter 2, only researchers who are willing to use inferential statistics may find it possible to publish articles in such journals or to secure research funding from such sources. In such cases, it is often possible and advantageous to use inferential statistics to augment, not replace, graphic analyses.

2. *Do not transform or statistically analyze data before plotting them.* By inspecting the plotted (graphed) data, one can readily identify similarities or differences in behavior across conditions and any unusual observations. If one or a few observations appear to be anomalous, one should determine whether there is an obvious cause that might make it appropriate to exclude the data from analysis (e.g., the apparatus malfunctioned), as discussed in Chapter 7.

As a rule, one should be cautious about presenting transformed data that lead to conclusions that would not be supported by the raw data. Consider, for instance, the data depicted in Figure 8.3. This figure shows the results of a hypothetical study that used an A–B–A–B design with a single subject to evaluate the effects of a contingency-management program for treating obesity in a patient with Prader-Willi syndrome (this condition is associated with severe obesity). Each condition was in effect

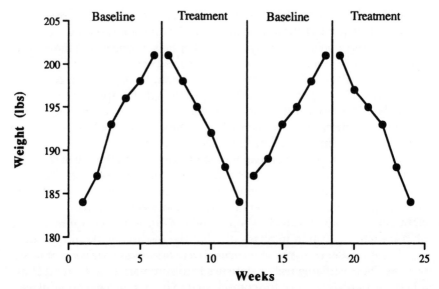

FIGURE 8.3. Hypothetical raw data showing the weight of a person with Prader-Willi syndrome across time under treatment (contingency-management) and no treatment (baseline) conditions. The person was weighed once each week.

for 6 weeks, and the patient was weighed weekly. The patient's weight increased progressively from the first to the last week under each of the A phases and decreased progressively under both of the B conditions. These trends are readily apparent upon visual inspection and suggest that the contingency-management program was effective in producing weight loss.

Figure 8.4 shows the mean weight of the patient across all four phases. The means are very similar; there is no suggestion of a treatment effect. This conclusion, which differs from that suggested by the raw data, would not be supported by any reasonable person aware of the raw data.

In some situations, an orderly effect of an independent variable will be evident when data are summarized in a particular way, but not when the raw data are considered alone. So long as the orderliness is not an artifact of the summarization process, such outcomes are interesting and potentially important. There are conventions for summarizing data in particular research areas, but one should not adhere blindly to them. In truth, the only way to determine the best way to summarize and analyze data is to test several alternatives. In testing alternatives, be aware that the more data are subjected to transformation and summarization, the greater

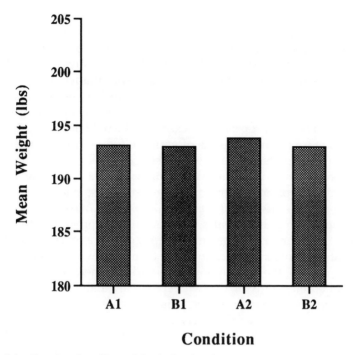

Condition

FIGURE 8.4. Raw data from Figure 8.3 calculated as the mean weight under each condition. Note that no treatment effect is evident in this figure, whereas a clear effect is evident in Figure 8.3.

is the risk of losing track of the actual behavior of individual subjects, which should be the focus of most behavior analytic research. As Johnston and Pennypacker (1993b, p. 298) emphasize:

> [T]he more an analytical procedure changes the investigator's picture of the subject's behavior as it actually happened, the greater the risk that the analytical procedure may exert more control over interpretations than the data. This means, for example, that a descriptive statistical procedure that reduces the data to a mathematical function describing a particular characteristic should be chosen with greater caution than a graphical technique that preserves the ebb and flow of responding from session to session. Although the statistical procedure may show something important that would otherwise be difficult to see, it will greatly change the stimuli that are controlling the investigator's reactions. In one case, this might result in a better summary description of the data, but in another instance, it might only distract the investigator from seeing features that a different view of the data would reveal. This same caution can be given about a graphical summary that reduces detail in the interest of showing a more general feature.

3. *Do not use inferential statistics that fail to emphasize the important characteristics of the plotted data.* By graphing the results of an experiment, one can detect noteworthy patterns in the data. These patterns should not be ignored when the data are subjected to further analysis. Consider the hypothetical study of weight loss in a patient with Prader-Willi syndrome that was discussed in the preceding section. The form of the data plot in Figure 8.3 tells us that an inferential test based on the means of the four phases would be inappropriate. For example, it would not be acceptable to use a repeated-measures analysis of variance followed by planned comparisons to analyze these data. It would be acceptable (albeit unnecessary, save for political reasons), however, to analyze these data using a regression model that considered trends in the data across phases.

4. *Do not use statistics that you don't understand.* Unless they are exceptionally well trained in statistics, and few are, neophyte researchers are likely to be familiar with only a limited range of tests and to be uncertain of appropriate instances in which to apply them, the exact kind of information they provide, the assumptions that underlie them, and their limitations. Nonetheless, given the widespread availability of computers equipped with statistical software, it is easy enough for even neophyte researchers to apply a wide variety of statistical tests to their data. As Huitema (1986b, p. 220) pointed out, this facility leads to potential problems:

> Until recently researchers tended to avoid computing certain complex analyses because they did not understand the associated mathematics or the need for such procedures. It has probably been a good thing that the authors of advanced statistics books routinely wander off into hyperspace never to be understood again. But mathematical intimidation and tedious number crunching sessions are no longer effective in preventing the computation of esoteric statistics. Computer software is now available to allow us to compute easily the most complex analysis imaginable. But, as any Apple should know, to compute is not to understand.

Using statistics without understanding can lead a researcher to erroneous conclusions. As a general rule, one should be able to relate statistical results to particular features of the plotted data. Assume, for example, that a researcher applies a regression discontinuity analysis to the data shown in Figure 8.3. The analysis reveals that the statistics associated with the changes in slope from condition A_1 to condition B_1, from condition B_1 to A_2, and from A_2 to B_2 are significant. This finding simply means that there is a change in degree of trending in the data across phases—but this feature of the data is already apparent in the graph.

A researcher who desires to use inferential statistics but is less than expert in their use has three choices: Do what others have done, seek expert advice, or become well trained. Doing what others have done is relatively easy, but may result in the persistence of inappropriate practices. For example, as discussed previously, regardless of the inferential statistic calculated, testing the null hypothesis tells us only the probability that the data could have arisen if the null hypothesis were true. It does not tell us the probability that the null hypothesis is true, given the data (Cohen, 1994). Many authors fail to make this distinction, and errors in interpreting tests of the null hypothesis persist, as several authors have emphasized (e.g., Cohen, 1994; Dawes, 1988; Morrison & Henkel, 1970). As a second example, Rosnow and Rosenthal (1989a) found that of 191 research articles that reported an interaction effect with analysis of variance, only 1% interpreted the interaction in an unequivocally correct manner. These examples illustrate an important point: *The simple fact that a statistical test has been repeatedly used in a particular context, or consistently interpreted in a certain way, does not necessarily mean that the usage or interpretation is correct.* One may be able to publish manuscripts that contain inappropriate, but precedented, statistical analyses (which go unnoticed by referees who are less than expert in statistics), but plodding along behind precedent is not the best way to develop a science of behavior.

Seeking the assistance of a trained statistician is a better tactic than blindly following precedent. There may be problems, however, in ascertaining who is adequately trained and in securing their assistance. Moreover, seemingly well-trained statisticians may disagree on the appropriateness of a particular analysis in a given situation. For instance, as mentioned previously, there is no consensus concerning the application of time-series analyses to data from individual subjects (see Huitema, 1986a, 1988). A researcher who lacks statistical knowledge is forced to agree blindly with the position of his or her selected expert and will not be able to defend the chosen practice if it is challenged.

The best course for any researcher who intends to use descriptive or inferential statistics is to become expert regarding them. Interestingly, one of the arguments advanced by Michael (1974) against the use of statistics in behavior analysis was an opportunity–cost argument. With respect to statistics, he asserted (p. 647), "The scientist must spend some time learning about them, time he or she might be spending in other activities relevant to his [or her] subject matter." Moreover, he added (p. 647), "Statistics courses displace other topics from the curriculum."

There may be activities that benefit an aspiring researcher more than studying statistics, but opportunity-cost is not sufficient justification for ignorance. Science is fundamentally quantitative. Even if one takes the hard-nosed stance that statistics have no useful role in behavior analysis, they are widely used in other areas. Good researchers in behavior analysis often make us of findings from other areas and must adequately understand statistics to be a good consumer of these results. In using statistics to analyze data, or in interpreting the results of studies that use statistics, a little knowledge is indeed a dangerous thing.

5. *Do not use statistics in violation of their assumptions.* Inferential statistical procedures involve fitting a model to the data. A model is an equation that attributes the total amount of variation in a data set to its component sources. Some models explain variation in terms of the mean of each phase and an error component. Other models include terms for trends in the data. Yet others include means and trends in their explanation. Each inferential test and its underlying model are associated with certain assumptions concerning the nature of the data being analyzed. In general, these assumptions must be met in order for the results of a statistical test to be accurate and valid. For example, when analysis of variance (ANOVA) is used, it is assumed that there is homogeneity of variance in the populations being sampled. This assumption is not crucially important when the number of subjects in each group is similar, but it is important when group size differs markedly. Tests (e.g., simple F-ratios) are available to determine whether the assumption of homogeneity of variance is upheld in a particular data set and, whenever there are sizable differences in the number of observations under different conditions, should be employed prior to conducting an ANOVA.

6. *Do not confuse statistical significance with clinical or theoretical importance.* The clinical significance of the effects produced in an applied study cannot be determined through statistical (or graphic) analysis alone, but must be assessed through the application of social validation procedures. Moreover, as repeatedly emphasized in this chapter, statistically significant results may be very small. Unfortunately, many investigators include discussion of statistically significant and nonsignificant results without mention of the size of the effect. Graphed data accompanied by some descriptive statistics inform the reader about the relative effect of an intervention, allowing both the research consumer and the investigator to judge whether the effects are of any practical value. If the mean difference, for example, is trivial, there is little justification for computing inferential statistical tests on the data (Huitema, 1986b).

Even if effects are large, one should never forget that a statistically significant result can be utterly meaningless. Consider, for example, a study that examined the level of employee readiness to participate in management activities (Turner, 1991). The study used an A–B design. Initially (in the A phase), 25 of 30 potential participants completed a Likert-scale questionnaire concerning their perception of actual and ideal management styles within a work unit. The intervention involved a series of organizational development activities. After being exposed to the intervention (in the B phase), 27 of 30 potential participants completed the same questionnaire. The questionnaire quantified 18 dimensions of management style (e.g., direction of information flow, whether subordinates are involved in work-related decisions). Comparisons of mean group scores from the A and B phases via t-tests were associated with p values of less than 0.05 in 5 of the 18 areas. What conclusions, if any, do these data support?

None with confidence. As discussed in Chapter 5, the design of the study, an A–B, is too weak to support strong conclusions. Moreover, the statistical analysis is inappropriate; 18 comparisons, each at the nominal 0.05 α level, yield an experiment-wide α of 0.9. Given this result, it would be a great surprise if at least one of the t-tests did not yield a significant p value regardless of whether or not the population means actually differed.

Obviously, as emphasized early in this chapter, the conclusions that people reach regarding the outcome of an experiment depend on the problem under investigation, the procedures used to investigate that problem, the size and consistency of the effects obtained, and the relation of these effects to those reported in similar investigations. Statistical analysis can provide information about *how* data differ across conditions, but they can never verify *why* they differ. Assigning observed effects to an independent variable requires consideration of the entire experimental context, not the data alone. This caution holds regardless of whether the data are analyzed graphically or through inferential statistics.

7. *Do not report statistical results without adequately describing statistical procedures.* An essential feature in the description of any study is to provide sufficient information to allow other people to understand the study and, if they wish, to replicate its procedures. This convention extends to data analysis techniques. That is, one should clearly describe any statistical procedures that were used, the specific data to which they were applied, and the exact results of those procedures. For instance, serious information is lost if the only information provided about an ANOVA is the obtained F-ratio and the corresponding p value. In this case, research consumers would benefit from also seeing the mean scores obtained under each condition and the amount of variability obtained within and between conditions. This information would assist the consumer in determining the size and importance of the obtained effects.

8. *Do not wait until a study is completed to select an appropriate statistical analysis.* A fairly common occurrence in certain research areas, among them behavioral pharmacology, is for referees of journal submissions to request a statistical analysis of data that initially were analyzed only graphically. Such a reaction may pose serious problems for the scientist

whose work is being evaluated, for there may be no test available that provides the desired information given the nature and extent of the data at hand.

As a rule, the researcher should determine before beginning a study whether there is any chance that the data from that study will be analyzed via inferential statistics. If there is, the researcher must make decisions concerning (1) the specific comparisons of interest, (2) the tests that will be used to make these comparisons, and (3) the legitimate interpretation of obtained results. She or he must also estimate the number and nature of the observations required to make the comparisons meaningful from a statistical perspective. If all these decisions are made appropriately, and if the experiment is carried out in a manner consistent with those decisions, inferential statistics can be legitimately applied to the data. Whether the benefits of having an appropriate statistical analysis available offset the costs of a fixed a priori approach to research design is a complex question that can be answered only case by case. For most behavior analytic research, the question is irrelevant, for neither the producers nor the consumers of that research require inferential statistics.

Summary and Conclusions

In a general discussion of data analysis techniques, Cohen (1990) offered two useful dictums: "Less is more" and "Simple is better." Beginning researchers will do well to heed this advice. In fact, when it comes to the use of statistics in analyzing behavior analytic data, it is tempting to go so far as to suggest: "None is best." There are exceptions, however, and both descriptive and inferential statistics can play a legitimate role in a science of behavior. That role is largely limited to revealing order in large and complex sets of data that cannot be easily analyzed via visual inspection alone.

It is easy to err in using statistics. The probability of doing so can be minimized by careful planning. The choice of whether or not to use inferential statistics should be made when an experiment is designed, not as a last-ditch attempt to salvage something of value from apparent chaos. Statistical procedures are not substitutes for good experimental control and should not be used in an attempt to impose order on highly variable data.

When a statistical procedure is desirable, it is imperative that the correct test be used and that its results be carefully described and appropriately interpreted. Statistical significance alone does not render experimental results theoretically or practically important. Moreover, there is no assurance that the results of any single experiment can be reproduced, no matter whether the results were detected via inferential statistics or graphic analysis. In science, important results must be reproducible, and the reproducibility of findings can be determined only through attempts to replicate them. Given this proviso, it is unwise to make too much of the results of any single experiment, be it yours or someone else's. *As a rule, be cautious about drawing broad conclusions based on single studies, no matter how the studies are conducted or analyzed.*

CHAPTER 9

Disseminating Research Findings

To be of benefit to other scientists and practitioners, a researcher's findings must be shared. Formal dissemination of findings culminates a research project and is accomplished through conference presentations and publications. This chapter considers what is involved in making a conference presentation and in publishing a journal article describing a research study. Although publications of other sorts (e.g., books, book chapters, review articles, research monographs) are important in integrating and interpreting research findings, they will not be considered here. Such writings require skills similar to those necessary to publish research articles, which are the foundation of the scientific literature.

Making a Conference Presentation

Scientific conferences provide an opportunity for people with common interests to share information and ideas, and to socialize. They vary in size from local meetings attended by a handful of participants to international conventions with an audience of thousands. Some large conferences of interest to behavior analysts are those of the Association for Behavior Analysis (ABA), the American Association for Behavior Therapy (AABT), the Psychonomic Society, and the American Psychological Association (APA).

Choosing an appropriate conference at which to present one's findings can be difficult. Information about the topics characteristically considered at a particular conference is available from the organizers; especially useful are conference proceedings from previous years. If past presentations are similar to those you intend to give, the conference merits your consideration—assuming, of course, that you want to attend the conference and have the wherewithal to do so. Although there are ways to cut costs (e.g., sharing a room with friends, staying at a cheap nearby hotel instead of the conference center, driving instead of flying, taking food instead of eating out), attending a conference can be expensive as well as time-consuming. Before submitting anything to a conference, be sure that you (or someone else who can make the presentation) will be able to attend it. Actual attendance requires action well in advance of the conference to ensure that lodging is secured, travel is arranged, and registration is accomplished. *You will greatly increase your likelihood of successfully participating in a convention if you exercise forethought and then take prompt action.*

To minimize potential problems, beginning researchers (such as undergraduate students) might want to attend a small local conference prior to visiting a large international meeting and make their initial presentation at that conference. They also are well advised to attend at least one convention without presenting anything, simply observing the actions of established professionals. *Presentation techniques that have succeeded historically are likely to be successful, and it is wise to mimic them.*

In general, four kinds of presentations are made at conventions: posters, addresses, invited addresses, and symposia. Specific written guidelines for making each kind of presentation are provided by the organizers of particular conventions, and those guidelines must be followed to the letter. Almost all conventions require submission of a written document describing a proposed presentation. A surprising number of submissions to conferences are rejected because the form is incomplete, inaccurate, unreadable, or incomprehensible, or because it is received after the application deadline, which typically is well in advance of the conference. For example, an ABA conference was held in late May, 1994, but completed submission forms were due by November 15, 1993, over 6 months prior to the meeting. *To increase the likelihood that a conference presentation will be accepted, prepare the submission form carefully and submit it well in advance of the deadline.* When the form is received, it will be evaluated and the submitter will be notified in writing whether or not his or her presentation is accepted. Some conferences accept almost all submissions; others are more restrictive.

Because there is a considerable delay between the deadline for applying to present at a convention and the time of the actual presentation, it may be tempting to submit an application based on ongoing, but incomplete, work. Doing so is a risky business that cannot be recommended. Better to wait for next year's convention than to act prematurely and propose a presentation that is based to a significant degree on data that have not been collected. They may never be.

Although one must know the contents of a conference presentation before submitting an application form, one usually begins final preparation of a presentation only when the presentation is accepted by the conference organizers. The following sections present general guidelines for preparing various types of conference presentations.

Poster Presentations

Poster sessions once were rare at conventions, but in recent years they have become commonplace. In these sessions, groups of thematically related posters are displayed for a relatively long time (e.g., an hour or two), usually in a large hall. The posters are arranged in long rows, and conference attendees file past them, scanning for those of interest, which they approach and read carefully. One or more preparers of each poster typically stand by their work and are available to discuss it with readers should the occasion arise, which it frequently does.

The attendees at a poster session can be exposed to a great number of studies in a short period of time, and they are not forced to give careful attention to studies that do not interest them. For presenters, posters afford a relatively easy way to disseminate the results of their research and to secure comments on it without the burden of oral presentation or extensive writing, which some scientists dislike or do poorly. Presenting a poster is a good place for neophyte researchers to begin disseminating their findings.

Specific instructions for poster presentations are supplied by all major conferences that use this format. These instructions characteristically specify the size of finished posters and may dictate acceptable fonts (sizes and styles of lettering) and colors as well. Most conventions provide large posterboards mounted on easels. Presenters typically bring their posters in appropriate bits and pieces and attach them to the posterboards with pins or tape. An alternative method is to prepare the poster in its entirety or in a few large pieces, which characteristically produces a more aesthetically pleasing poster, but one that is hard to transport.

In preparing and presenting any poster, it is useful to follow five guidelines:

1. *Make the poster eye-catching.* The title should be short and informative, so that readers can discern your topic at a glance. Colors should be coordinated, and the poster should be laid out carefully.

2. *Make the poster easy to read.* Ensure that letters are large enough to read and that the reader's eye is directed from section to section in appropriate order. Many people organize posters around the five sections characteristic of research articles: introduction, methods, results, discussion,

and references. An abstract may also be included—usually as the first element, because it serves as a digest for the potentially interested reader. Although this format works and will be familiar to most readers, it is not sacred.

3. *Cut detail to a minimum.* Posters are not an appropriate medium for presenting a study in its entirety; that is the role of the journal article. Posters offer an excellent mechanism for the summary interchange of information with minimal effort, an advantage that is maintained only if the reader can easily grasp the essentials of a study. Preparing such a summary requires paring away superfluous details. Extensive literature citations are inappropriate for posters; mention only key articles directly relevant to the purpose and findings of the work presented. Reduce procedural descriptions to general summaries, and present findings in the same way. One or two simple, self-explanatory figures are enough for most posters. Keep conclusions simple; don't equivocate. Avoid technical terms wherever possible, for much of your potential audience consists of non-specialists. Good posters are painted with broad, bright brush strokes.

4. *Have details available for readers who desire them.* It is appropriate to note on the poster that further details about the study can be obtained by talking with the presenter, who should remain by the poster. It is also appropriate, although not absolutely necessary, to prepare a manuscript that describes the study in detail and to make copies available for interested parties. This strategy generally is used when the poster describes a study that has been submitted to a journal for review or is currently in press.

5. *Have copies of a written abstract available.* This document should summarize the study and provide a means of contacting the people who conducted it (e.g., the name, address, and phone number of a contact person). Giving readers a take-away document increases the likelihood that their subsequent behavior will be influenced by the study reported in a poster. Lacking such a document, readers are likely to read your study and then quickly forget it.

Oral Presentations

Oral conference presentations include addresses, invited addresses, and symposia. They are sufficiently alike to be discussed together, differing primarily in that invited addresses and symposia are specifically solicited, whereas general addresses are not.

There is no single best tack to take for oral presentations. Some able conference speakers are droll, others dry. Nonetheless, all good speakers behave similarly in some regards, as summarized in the following rules for oral presentations:

1. *Have something to say.* The worst oral presentations are desultory, without obvious focus or conclusions. The best are tightly organized, easy to follow, and unambiguous. After hearing an oral presentation, members of the audience should be able to state clearly (1) the topic discussed, (2) the importance of that topic, (3) the methods used by the presenter to examine that topic, (4) the researcher's findings and their relation to the topic, and (5) the researcher's conclusions concerning the topic. A good oral presentation based on data is possible only if those data are sufficient to support meaningful conclusions. In many cases, the results of a single study do not merit an oral presentation, but are better shared via a poster.

One useful tip for increasing the likelihood of a quality presentation is to emphasize only two or three take-home messages. Others are to (1) build in redundancy to emphasize those messages, (2) use short and simple sentences, (3) match the level of the presentation to the audience (err on the side of simplicity), and (4) ensure that transitions from topic to topic are logical and smooth (Critchfield, 1993). Digressions, even if clearly indicated as such, usually detract from a presentation and should be included only with good reason. A bit of humor, well done, enhances a presentation—just be sure it's inoffensive.

2. *Don't read.* With very few exceptions, oral presentations delivered straight from a prepared text are boring and uninformative. Many speakers do produce a full written text of their oral presentations, and such a text is useful in that (1) preparing it helps the presenter prepare to speak before an audience, (2) it provides a useful crutch should the oral presentation go awry, and (3) it helps keep the speaker on task and away from digressions (Critchfield, 1993). Copies of the text can also be a useful resource for interested members of the audience, although the copies should be distributed after, not before, the oral presentation. If they are distributed beforehand, members of the audience are likely to read, not listen, during the presentation.

Although it is reasonable to prepare a verbatim transcript of an oral presentation, most good speakers use a simple outline, not a narrative, as an aid in presenting. Sidebar 9-1 presents a sample presentation outline. In this outline, **bold** type is used to remind the speaker to emphasize certain points and to present figures. Like a narrative, a skeleton outline helps keep the speaker on task and provides a crutch in case the speaker falters. With the addition of timing cues (like those in the right margin in Exhibit 9-1), an outline can also help a speaker pace the presentation appropriately, speeding up or slowing down as necessary.

3. *Use audiovisuals wisely.* Good audiovisual aids potentially serve

Sidebar 9-1
Sample Presentation Outline

The outline reproduced below has been used on several occasions by Alan Poling in presenting an overview of psychotropic drug effects in people with mental retardation. This presentation lasts about 45 minutes, which allows sufficient time for questions (and miscues), given a 60-minute time slot. Times listed in the right margin indicate when discussion of the indicated figure (or topic) should end.

1. Opening remarks
   A. I'm pleased and honored to speak with this audience.
   B. I'm also somewhat embarrassed, because I've been doing this same talk for almost 10 years—many of you could give it, too.
   C. I'm going to summarize findings and the current state of the field, then offer some speculations as to what is needed. More detail concerning findings in two sources **(Figure 1)**.   **5 MIN**
2. Like other people, those who are mentally retarded have some behavioral problems:
   A. Do things they shouldn't do.
   B. Fail to do things they should do.
   C. Do things at the wrong time or in the wrong place.
3. Caregivers attempt to change the behavior of mentally retarded people to improve their lives:
   A. Many different strategies are available.
   B. Two common ones involve medications and behavior management procedures.
4. Drugs are commonly used **(Figures 2-6)**. Note John Jacobson (1988, *Research in DD*) found in a study of 35,000 people that young and middle-aged adults were most likely to receive psychotropic drugs, that receipt rates increased with severity of retardation, with increasing restrictiveness of residential setting, with rated severity of problem behaviors, and with concurrent mental illness.   **10 MIN**
   A. Used because:
      1. They are effective with mentally ill people **(Figure 7)**.
      2. They are relatively easy to use.
      3. They appeared to be effective with mentally retarded people.
5. Nonetheless, in the last decade skepticism has grown concerning the use of psychotropic medications with mentally retarded people:

A. Thomas Greiner as early as 1958 expressed skepticism; quote from *AJMD* paper:

"In the years to come, the retarded may claim an all-time record, of having the greatest variety and largest tonnage of chemical agents shoveled into them. Sensible adult patients will usually balk when a drug is causing symptoms, but the very young and the very old are forced to take drugs, can't complain or stop on toxic symptoms, may not even connect them with the drug. The mentally deficient of any size or age cannot protect themselves either, and they also merit special care to avoid toxic doses."   **15 MIN**

B. Reasons for current skepticism:
   1. Drugs are nonselective and potentially harmful (**Figure 8, Figure 9**).
   2. Drugs are not generally useful (**Figure 10, Figure 11**, CPZ and Haloperidol).   **20 MIN**
   3. Many studies are methodologically weak (**Figure 12**).   **25 MIN**
6. It is now widely recognized that, historically, neuroleptic drugs were used too often, and too high doses. Courts have now set limits on how these drugs are used in state facilities (**Figure 13**).   **35 MIN**
   A. Nonetheless, drugs sometimes are very useful (**Figure 14**).
   B. Usefulness is apt to be greatest in combination with appropriate contingency management (**Figure 15, Figure 16**).   **40 MIN**
7. It is now widely recognized that behavior analysts have a major role to play in assessing drug effects, in developing adjuncts to pharmacological treatments, and in developing alternatives to pharmacological treatments (**Figure 17**).
8. That's what's known, now what's needed?
   A. Additional research (**Figure 18**):   **40 MIN**
      1. Note Thompson's findings.
      2. Note Sprague's findings.
      3. Note Harder, Kalachnik, Jensen, & Feltz RDD study of drug use as it relates to successful community placement (from Cambridge).
   B. Integration of behavioral and pharmacological treatments.
   C. No more poor-quality research. Quote Greiner:
      "If your aim is helping the retarded, I urge you to avoid the casual clinical trial of drugs. Make them good trials, or don't make them at all."
      To which I'd add, "AMEN."
9. Thank audience, entertain questions.   **50 MIN**

three important functions (Critchfield, 1993). First, they allow rapid presentation of complex information (e.g., the details of an experimental procedure or the results of a study) that would otherwise necessitate a lengthy and complex verbal description. Second, they help inform the audience of the structure of a presentation and its main points. Third, they increase audience attention.

A picture is worth a thousand words—but those words are likely to be curses from the viewers if the picture is an unintelligible figure. Regardless of the medium (slides, transparencies, videotape), good visual presentations have three features: (1) they are germane to the topic of discussion, (2) they are easily readable from everywhere in the room, and (3) they are self-explanatory. In general, although there are no firm rules for determining the appropriate number of graphics for a given oral presentation, use significantly more than in written presentations (posters, journal articles). For most presentations, one figure every 2 to 5 minutes is about right. As a rule, graphics should be removed as soon as the presenter has discussed them, and there should be no long gaps between audiovisual presentations (Critchfield, 1993). Be sure that apparatus suitable for the kinds of audiovisuals that you intend to use is available at the conference, and familiarize yourself with that apparatus before your presentation (it's terribly easy to reverse slides). Finally, recognize that the apparatus may malfunction before or during your presentation and be prepared to make a creditable presentation without your audiovisuals.

4. *Practice.* Making effective oral presentations is a skill that people acquire; there are no born orators. Observing established speakers facilitates this process, but there is no substitute for actual practice. Before delivering a conference presentation, practice delivery before a mirror, as well as in front of any audiences that can be arranged. Colleagues, instructors, and classmates are often willing to listen to an informal presentation and to provide sympathetic feedback, perhaps in exchange for the same favor when their turn comes. Audiovisual tapes allow speakers to monitor their own performance and to improve it prior to making a formal presentation. Finally, formal classes in public speaking are useful in refining skills, although scientists rarely attend them.

5. *Watch the clock.* Good presentations end on schedule, without rushing. To be timely, err on the safe side by making sure that your presentation, delivered without interruption, will take about 75–80% of the allotted time (e.g., 15 minutes for a 20-minute time slot). The remaining time will probably be used up by unforeseen events, such as a late start, unplanned question, or audiovisual malfunction. If there is no such interruption, it is no sin to end early. In fact, doing so may win the applause of both your audience and your fellow speakers.

The best way to learn pacing is through practice. Timing cues, especially those linked to the presentation of particular audiovisuals, are likely to prove helpful at the time of the actual presentation. For example, a speaker who knows that a figure that should have been presented 10 minutes into a 20-minute presentation was not presented until the 12-minute mark (a stopwatch placed on the podium in easy sight of the speaker is handy for timing such things) obviously knows that she or he must increase the pace. The pace can be quickened by omitting some material initially planned for discussion or by discussing topics in less detail than planned. Either tack is reasonable, but neither can be easily accomplished if the presenter is delivering the presentation verbatim from a detailed narrative. One advantage of working from a skeleton outline is that it encourages the kind of fluid style that allows speakers to modify their delivery as time and audience demands.

6. *Don't give up.* Not all problems can be foreseen; things sometimes just go wrong. Sessions start late and are poorly attended, microphones and slide projectors give up the ghost, members of the audience ask rude and foolish questions. On such occasions, the best that can be expected of any speaker is grace under fire. Press on, return to schedule as soon as possible, and make whatever adjustments are necessary to complete the presentation.

If you err, perhaps by failing to make an important point or by speaking unclearly, note the error, apologize briefly, make corrections, and keep going. Audiences neither demand nor expect perfection; an error or two won't seriously tarnish an otherwise good presentation.

Of course, errors aren't desirable, and the likelihood of their occurrence goes down with practice. Learn from experience, keep making presentations, and endeavor to make each one better than the one before. If they can be arranged, audiovisual tapes are useful in helping speakers to monitor and improve their performance. Detailed notes recording strengths and weakness of each presentation, including feedback from members of the audience, serve a similar function, albeit less objectively. Systematic effort is required to improve as a public speaker. Fortunately, the rewards of that effort are readily apparent, both to speakers and to their audiences.

Publishing a Journal Article

Posters and oral presentations allow for rapid dissemination of research findings and are valuable for that reason. But it is the journal article that serves as the primary archival repository of research findings. Research articles constitute the empirical foundation of all sciences, and scientists are judged, in large part, according to the quality and quantity of the research articles they publish.

The process of publishing can be envisioned as a series of steps:

1. *A study is designed and conducted.*
2. *The results of the study are analyzed.* Prior chapters considered issues related to designing and conducting studies and to analyzing data.
3. *A decision is made to prepare a manuscript describing the study and to submit that manuscript to a particular journal for possible publication.* Many journals, including those listed in Table 9.1, frequently publish behavior analytic research. These journals differ significantly with respect to the range of topics they address, the methodological requirements they impose, and their willingness to publish studies that are limited in scope, significance, or methodological rigor. All journals provide guidelines for contributors, and these guidelines help to delineate requirements for publications in the respective journals. Nonetheless, one must read extensively in a given journal to determine whether that journal is an appropriate outlet for a given study. Consulting with people who have recently published in a particular journal is another way of evaluating whether it is a feasible outlet for a particular study. Finally, there is nothing wrong with calling a journal's editor and discussing whether a given study is of interest to that person. Although editors obviously cannot commit to

TABLE 9.1. Journals That Frequently Publish Behavior Analytic Research Articles

Outlets for basic research
  Animal Learning and Behavior
  Behavioural Processes
  Journal of Experimental Psychology: Animal Behavior Processes
  Journal of the Experimental Analysis of Behavior
  Psychological Record
Outlets for applied research
  Analysis of Verbal Behavior
  Behavior Modification
  Behaviour Research and Therapy
  Journal of Applied Behavior Analysis
  Journal of Organizational Behavior Management
Outlets for behavioral pharmacology research
  Behavioural Pharmacology
  Experimental and Clinical Psychopharmacology
  Journal of Pharmacology and Experimental Therapeutics
  Pharmacology, Biochemistry and Behavior
  Psychopharmacology

publishing something they have never seen, they can indicate whether or not a study appears to be appropriate for their journal.

4. *The manuscript is prepared and submitted to the journal's editor.* Manuscripts should be prepared and submitted in exact accordance with the directions provided in the journal. Sidebar 9-2 presents an example of such directions, specifically, guidelines for the preparation of submissions to the journal *Pharmacology, Biochemistry and Behavior*. Manuscript preparation is discussed further below.

A submitted manuscript should be accompanied by a brief letter, which usually (1) requests that the manuscript (specify title and authors) be evaluated for possible publication in the journal (as a particular kind of article, if the journal uses different categories), (2) indicates that the manuscript presents new findings and is not currently under review elsewhere, and (3) specifies the name, address, and phone number (as well as fax and E-mail numbers) of the contact author. Some authors also indicate in their submission letter that the research described was approved by an appropriate human subjects institutional review board or institution animal care and use committee. Sidebar 9-3 presents an example of a submission letter, one that we used in submitting a study from our laboratory to the editor of *Pharmacology, Biochemistry and Behavior*.

The editors of most journals send a card or letter of acknowledgment to the submitting author when a manuscript is received. The acknowledgment typically indicates the identification number assigned to the manuscript and may provide an approximate timeline for the review process. The name of the associate editor to whom a manuscript is assigned (if any) is also characteristically indicated. Some journals send transfer of copyright forms to the author at the time an article is submitted; others wait until the article is accepted for publication. If the submission is lacking in some regard (e.g., no glossy figures are provided), this lack may be indicated in the initial correspondence. It is the author's responsibility to respond to such problems immediately.

5. *Copies of the manuscript are distributed to referees by the editor.* Peer review is the process that ensures quality control for research articles, and referees are the people who do the reviewing. Depending on the journal, two to five (or more) established professionals evaluate each manuscript submitted. Journals that use associate editors, such as the *Journal of Applied Behavior Analysis* and the *Journal of the Experimental Analysis of Behavior*, add a step to the review process. For such journals, manuscripts are routed from the editor to an associate editor, who sends them to referees.

6. *The manuscript is evaluated by the referees, who make recommendations to the editor.* Journals characteristically provide instructions to referees; an example of such instructions—those sent to *Pharmacology, Biochemistry and*

Instructions to Authors

The following instructions are for authors submitting articles for possible publication in the journal *Pharmacology, Biochemistry and Behavior*. The instructions appear on the inside back cover of the journal.

Papers submitted for publication with a European country of origin should be sent to Professor Sandra E. File, Psychopharmacology Research Unit, UMDS Division of Pharmacology, Guy's Hospital, P.O. Box 3448, London SE1 9QH, UK; FAX: 44-(0)71-955-4627. Papers from elsewhere should be directed to Professor Matthew J. Wayner, Division of Life Sciences, The University of Texas at San Antonio, San Antonio, TX 78249-0662; FAX: (210) 691-4510.

The covering letter accompanying the manuscript must include a statement that the experimental protocol was approved by an Institutional Review Committee for the use of Human or Animal Subjects or that procedures are in compliance with at least the Declaration of Helsinki for human subjects, or the National Institutes of Health Guide for Care and Use of Laboratory Animals (Publication No. 85-23, revised 1985), the UK Animals Scientific Procedures Act 1986 or the European Communities Council Directive of 24 November 1986 (86/609/EEC). Manuscripts will be returned if there is sufficient evidence that these accepted procedures and good ethical standards have not been followed.

Style of Manuscript

General Form. (1) Manuscripts should be submitted in triplicate and should be typewritten, double-spaced with wide margins on good quality paper. If a word processor is utilized to prepare the manuscript, a letter quality printer must be used and computer-generated illustrations must be of the same quality as professional line drawings or they will not be accepted. (2) The title page should contain: title of paper; author(s); laboratory or institution of origin with city, state, zip code, and country; complete address for mailing proofs; a running head not to exceed 40 characters including spaces between words. (3) References, footnotes, and legends for illustrations should be typed on separate sheets, double-spaced. (4) Illustrations (unmounted photographs) should be identified on the reverse with figure number and author(s) name; when necessary the top should be clearly marked. (5) Each table should be typed on a separate sheet and double-spaced. (6) All dimensions and measurements must be specified in the metric system. Standard nomenclature, abbreviations and symbols, as specified by Royal Society Conference of Editors. Metrification in Scientific Journals, Am. Scient. 56:159–164; 1968, should be used throughout. (7) Italics should not be used for the purpose of emphasis.

Title. The title should not be longer than 85 characters, including spaces between words.

Length of Paper. The Editors insist upon clear, concise statement of facts and conclusions. Fragmentation of material into numerous short reports is discouraged.

Abstract. Each paper submitted must be accompanied by an abstract, which does not exceed 170 words and must be suitable for use by abstracting journals. Abstracts should be prepared as follows:

MYERS, R. D., C. MELCHIOR AND C. GISOLFI. *Feeding and body temperature: Changes produced by excess calcium ions …* PHARMACOL BIOCHEM BEHAV. Marked differences in extent of diffusion have been …

A list of 3–12 (or more) words or short phrases suitable for indexing terms should be typed at the bottom of the abstract page accompanying the manuscript. These terms will be printed with the paper at the end of the abstract.

Drugs. Proprietary (trademarked) names should be capitalized. The chemical name should precede the trade, popular name, or abbreviation of a drug the first time it occurs.

Footnotes. Title page footnotes should be numbered consecutively. If the senior author is not to receive reprint requests, a footnote should be given to designate to whom requests should be sent. Text footnotes should not be used; the material should be incorporated into the text. Table footnotes: see *Tables* (b).

References. Literature cited should be prepared according to the Numbered/Alphabetized style of the Council of Biology Editors. References should be cited by number, in parentheses, within the text (only one reference to a number) and listed in alphabetical order (double spaced) on a separate sheet at the end of the manuscript. *Do not recite names of authors within the text.* Journal citations in the reference list should contain the following: (a) surnames and initials of all authors (surname precedes initials); (b) title of article; (c) journal title abbreviated as listed in the *List of Journals Indexed in Index Medicus*; (d) volume, inclusive pages, and year. Example:

1. Banks, W. A.; Kastin, A. J. Peptides and the blood-brain barrier: Lipophilicity as a predictor of permeability. Brain Res. Bull. 15: 287–292; 1985.

Book references should be in the following order: author, title, city of publication, publisher, year, and pages. Examples:

1. Mello, N. K. Behavioral studies of alcoholism. In: Kissin, B.; Begleiter, H., eds. The biology of alcoholism, vol. 2. Physiology and behavior. New York: Plenum Press; 1972:219–291.
2. Myers, R. D. Handbook of drug and chemical stimulation of the brain. New York: Van Nostrand Reinhold Company; 1974.

Computer Disks. In order to speed publication and ensure accuracy, authors are requested to submit a computer disk containing the final version of their paper along with the final manuscript to the editorial office. Please observe the following criteria: 1. Specify what software was used, including which release (e.g., WordPerfect 4.0); 2. Specify what computer was used; 3. Include both text file and ASCII file on the disk; *(continued)*

Sidebar 9-2
(Continued)

4. The file should be single-spaced and should use the wrap-around end-of-line feature (i.e., no return at the end of the line). All textual elements should begin flush left, no paragraph indents. Place two returns after every element, such as title, headings, paragraphs, figure and table callouts, etc.; 5. Keep a backup disk for reference and safety.

Illustrations. (a) Prepare for use in a single column width whenever possible. (b) All drawings for reduction to a given size should be drawn and lettered to the same scale. (c) All illustrations should be referred to as figures and numbered in Arabic numerals. (d) Lettering should be done in India ink or other suitable material and must be proportionate to the size of the illustrations if it is to be legible after reduction. Lettering should be sized so that its smallest elements (subscripts or superscripts) will be readable when reduced. (e) When possible all lettering should be within the framework of the illustration; likewise the key to symbols should be on the face of the chart. The following standard symbols should be used as they are easily available to the printer: ○ ● △ ▲ □ ■ +. (f) Actual magnification of all photomicrographs should be given. Dimension scale should be indicated. (g) Sharply contrasting unmounted photographs of figures on glossy paper are required. (h) Illustrations should be submitted in black and white unless color reproduction is requested. Color prints should be submitted in actual size and authors will be responsible for the additional costs.

Tables. (a) Each table should have a brief heading; explanatory matter should be in footnotes, not as part of the title. (b) Table footnotes should

---

Behavior referees—is presented in Sidebar 9-4. In general, referees evaluate submissions along six dimensions: suitability for the journal, importance of topic, soundness of methods, clarity of results, reasonableness of conclusions, and quality of writing. Any ethical problems that may be present are also noted. Referees provide a narrative evaluation of the manuscript, as well as a summary recommendation concerning it. In general, referees (and editors) recommend one of four actions for the disposition of a manuscript: accept, accept pending revision, reject and resubmit, or reject.

7. *The editor makes a decision concerning whether to publish the manuscript and transmits that decision to the author.* Editors evaluate manuscripts themselves and also refer to referees' evaluations in making decisions concerning publication. An editor's decision concerning a manuscript is transmit-

be indicated in the body of the table in order of their appearance with the following symbols: * † ‡ § ¶ # **, etc. (c) Tables must not duplicate material in text or illustrations. (d) Vertical rules should be omitted. (e) Short or abbreviated column heads should be used. (f) Statistical measures of variation, SD, SE, etc., should be identified. (g) Analysis of variance tables should not be submitted, but significant F should be incorporated where appropriate within the text. The appropriate form for reporting F value is: $F(11, 20) = 3.05, p < 0.01$.

Formulas and Equations. Structural chemical formulas, process flow-diagrams, and complicated mathematical expressions should be kept to a minimum. Usually chemical formulas and flow-diagrams should be drawn in India ink for reproduction as line cuts.

All subscripts, superscripts, Greek letters, and unusual characters must be clearly identified.

Anesthesia. In describing surgical procedures on animals, the type and dosage of the anesthetic agent should be specified. Curarizing agents are not anesthetics; if these were used, evidence must be provided that anesthesia of suitable grade and duration was employed.

Proofs. Corrections to the proofs must be restricted to printer's errors only. Other than these, any other alterations will be charged to the author.

Reprints. Each author will receive with his galley proofs a reprint order form which must be completed and returned with the proofs. The senior author receives 25 reprints free. Authors of papers that significantly exceed average size (six printed pages) are requested to purchase 200 reprints of their article.

Copyright. Publications are copyrighted for the protection of the authors and the publisher. A Transfer of Copyright Agreement will be sent to the author who submits the manuscript. The form must be completed and returned to the publisher before the article can be published.

ted to the submitting author in a letter, which is usually accompanied by the referees' reports. Exhibit 9-5 presents the editorial decision and the referees' comments for the manuscript described in Sidebar 9-3. The reviews were generally favorable, and the editor's decision was to accept the manuscript for publication pending minor revision. In this case, the manuscript was not sent to the referees for a second review.

The times required for review—that is, the time from submission of a manuscript to receipt of an initial editorial decision—differs substantially for different journals and even for different manuscripts submitted to the same journal. The initial review typically takes 3 to 5 months. If there has

---

Sidebar 9-3
Sample Cover Letter Accompanying a Submitted Manuscript

The letter reproduced below accompanied the manuscript of an article describing research from our laboratory. We did not specify that the article was not submitted elsewhere, assuming that this was a given, or that the study was approved by our Institutional Animal Care and Use Committee, which should also be a given. In all correspondence described in these samples, the letters were typed on letterhead paper and included appropriate dates, addresses, and names and titles.

Date

Editor's name
Editor's address

Dear Dr. _____ :

Enclosed herewith are three copies of a manuscript, entitled "The effects of *d*-amphetamine and diazepam on schedule-induced defecation in rats." We authors would appreciate having the manuscript considered for possible publication in *Pharmacology, Biochemistry and Behavior.*

Sincerely,

Submitting author's name and title
Enclosures

---

been no response to your submission within 6 to 9 months, a brief query to the editor (perhaps by telephone) is appropriate.

8. *The author responds to the editor's decision.* Acceptance of a manuscript for publication typically necessitates no immediate response from the author. It is rare, however, to have an article accepted for publication in its initial form. More common editorial decisions are to accept an article pending appropriate revision or to reject the article in its present form with the recommendation that it be revised and resubmitted. Both of these decisions necessitate preparation of a revised manuscript. Strategies for preparing revisions are discussed later in the chapter.

The author can take three tacks concerning a rejected manuscript: Forget about publishing it, submit the manuscript in its present form to another journal, or revise the manuscript and submit it to another journal.

Exhibit 9-4
Sample Journal Referee's Report Form

A form similar to that reproduced below is given to people who review manuscripts submitted to *Pharmacology, Biochemistry and Behavior.*

Pharmacology, Biochemistry and Behavior

Referee Report
(Please print or type)

Title:
Author:
Do You Recommend Publication:
Comments or Criticisms to Help Author Improve the Manuscript:

The referee's written comments are made in this section, which covers most of the page.

Instructions to Referee

In your evaluation, it would be helpful if you could provide answers to the following questions: 1. Is the article appropriate for publication in terms of interest, originality, and editorial policy? 2. Are there undesirable redundancies in the manuscript, figures, tables, etc.? 3. Should any section be abbreviated? 4. Are there any errors of logic or fact—as distinct from points of scientific controversy or literary style—which should be corrected or clarified? Any additional material or comments to the Editor which are not to be transmitted to the author can be typed on plain white sheets. Please Return Original and First Carbon of Referee Report to the Editor.

Date: _____ Signature: _____

The tack best taken depends upon the content of the review, the journal from which it was received, and the author's candid evaluation of the manuscript. It is inappropriate and unethical to submit the same article to two or more journals at the same time; always secure reviews from different journals sequentially, not simultaneously.

9. *The manuscript is resubmitted and the review process is repeated.* Most published articles are revised and resubmitted to the same journal one or more times. Each revision may be submitted to the full review process, or it may be evaluated by the editor (or associate editor) alone. (The latter was the case for the manuscript considered in the exhibits.) When submitted,

Exhibit 9-5
Sample Editor's and Referees' Comments
on the Submitted Manuscript

Reproduced below are the editor's and the referees' comments on the article referred to in Sidebar 9-3. In the actual review, the editor's comment was on letterhead paper, with the appropriate date, name, and address, and the referees' comments were on separate sheets.

Editor's Comment

Date

Author's name
Author's address

Re: Manuscript #PBB9209, "The Effects of D-Amphetamine and Diazepam on Schedule-Induced Defecation in Rats"

Dear Dr. _____:
　　We are pleased to inform you that your manuscript has been accepted for publication with minor revision. Enclosed are copies of two referee reports. Please revise your manuscript to incorporate these suggestions where appropriate.
　　Please return your revised manuscript to us within two weeks in duplicate, along with a disk copy, as it will be published in the next issue available at the time of receipt.

Cordially yours,

Editor's name and title

Enclosures

cc: Referees

Referees' Comments

REFEREE 1

Do you recommend publication? Yes.

This is a straightforward, well-executed experiment that is clearly presented. I have only a few remarks that I think will lead to improvement of the paper. In what order were the drug doses given—random, ascending, etc.? It is difficult to accept the study by Sanger and Blackman cited as valid evidence that diazepam increases schedule-induced drinking. Later work has shown that diazepam, as well as the injection of other benzodiazepines, produces increases in water intake that is not schedule-induced. The difference they claim, then, between diazepam effects on schedule-induced drinking and schedule-induced defecation may consequently be questioned. This does not in any way detract from the paper; the authors should carefully evaluate what they want to infer here. There is a large literature on the effect of BZs and other drugs on fluid intake, but a relevant, specific reference is: Cooper, S. J. Benzodiazepine mechanisms and drinking in the water-deprived rat. *Neuropharmacology,* **21,** 775–780, 1982. Wylie has just published another paper on schedule-induced defecation that has just appeared in J. Exp. Anal. Behav. that the authors may wish to cite.

REFEREE 2

Do you recommend publication? Yes.

I recommend that this paper be accepted for publication in *Pharmacology, Biochemistry and Behavior.* It is an original study, adequately justified in terms of relevant literature, appropriately designed and with interesting results, all presented clearly and on a subject appropriate to the Journal.

I have some minor comments:

Page 5: Were the different drug dosages presented in a random order (and in a different order across rats)?

Page 7, line 4: Spelling of defecate

Page 7, line 14: Reference 1 should be 11

I am not entirely convinced that the discussion of general arousal from the bottom of page 7 to the end of the manuscript really earns its keep, although there is no doubt that the present data can be evaluated in the light of this general theory (and of course others). In the bottom paragraph of page 8, I found myself puzzling about the choice of the word "nonspecific": I am not sure that I understand what is meant by this, nor am I clear about the impact of the final sentence of that paragraph.

In general, however, as I have said, I am enthusiastic about the publication of this nice study.

each revision should be accompanied by a cover letter that clearly specifies the changes made in response to the preceding review. Sidebar 9-6 presents the letter that accompanied our revised manuscript.

The general rule for preparing revisions is to address all the referees' concerns that are reasonable and possible to address. If one is unwilling or unable to make a suggested change or to deal with an alleged problem, it is important to indicate the reason for not doing so. Occasionally, referees differ in their suggestions, and the editor does not make obvious recommendations for preparing a revision. In such cases, it may be helpful to call the editor for further guidance, although relatively few authors do so.

10. *The manuscript is published.* Eventually, if all goes well, the author is informed that the manuscript is accepted for publication. At that time, some journals request a disk containing the manuscript as a computer file that is used in typesetting. Others bypass this step and send the manuscript itself to typesetters. In either case, the typeset manuscript, termed a *galley proof,* is eventually returned to the author for proofreading, after which the article is published. The author must read and correct galley proofs both promptly and carefully. Perhaps the best way to check proofs is to have one person read from the edited typescript and another check to see that the same words and numbers appear on the galley. Specific symbols and conventions, outlined in the *Publication Manual of the American Psychological Association* (APA, 1994), commonly called the *APA Publication Manual,* are used in the editorial process.

When an article is accepted for publication, the author is given a chance to purchase reprints, which are sent some time after the article appears in print. Although a few journals provide for rapid publication, there is characteristically a considerable delay (months to years) between acceptance of an article for publication and its actual publication. This delay prevents the timely dissemination of important findings, a problem that can be partially corrected by presenting findings at conferences prior to their publication. Doing so is a common, and commendable, practice.

Writing a Journal Article

The best way to learn to write publishable articles is to work closely with a successful author. Although it is possible to provide useful general rules for writing articles, there is no substitute for a good mentor. Such a person will prove invaluable in determining whether a given study merits publication and where it might be published, as well as in actually preparing a manuscript for submission. In behavioral terms, a competent mentor models

Exhibit 9-6
Sample Cover Letter Accompanying the Revised Manuscript

The letter mailed with the revised manuscript appears below. Because the manuscript was evaluated favorably and the referees' comments were easy to address, the changes made in revision were straightforward and did not merit elaborate explanation.

Date

Editor's name
Editor's address

Dear Dr. _____:

Thank you for seeing to the prompt and helpful review of manuscript #PB9209, entitled "The effects of *d*-amphetamine and diazepam on schedule-induced defecation in rats." We authors are pleased to learn that a suitably revised version will be acceptable for publication in *Pharmacology, Biochemistry and Behavior*.

Enclosed herewith is a revised version of the manuscript, along with a disk containing it as a Microsoft Word document and a completed copyright agreement. We made the following changes in preparing the revision:

1. As both reviewers suggested, we now explain the sequence in which drug doses were administered (page 5, lines 21–22).
2. As Referee 1 suggested, we now point out that the increases in fluid intake produced by diazepam in previous studies may not be specific to schedule-induced drinking (page 3, lines 18–22).
3. As Referee 2 suggested, we have shortened and clarified our consideration of the possible role of general arousal in schedule-induced defecation (page 8, final paragraph).

We have also included the two references suggested by Referee 1 and corrected the two minor errors noted by Referee 2.

These changes, although relatively small, are important, and we thank the referees for suggesting them.

Sincerely,

Submitting author's name and title
Enclosures

appropriate scientific behaviors, and also provides rules and arranges contingencies that support those behaviors in students.

In the absence of a skilled advisor, the best advice is to mimic the characteristics of previously published articles. These characteristics become evident only upon close inspection; good scientific writers are well read, particularly with respect to the journals in which they regularly publish. They also use the language with skill; their writing is good. Good writing has three defining characteristics:

1. *Good writing is clear.* Clear writing is easy to understand and conveys information without ambiguity. Clarity is, in part, audience-dependent; what is clear for a trained specialist may be unclear for other people. A workable test for the clarity of part or all of a scientific manuscript is to have two or three appropriately trained people read it, then list the important points. If the writing is clear, there should be substantial agreement across readers in the points listed. Moreover, readers should agree that the writing is clear and easy to follow. If they do not, revision is probably needed.

2. *Good writing is concise.* The purpose of a scientific article is to inform readers, not impress them. Long sentences and ten-dollar words diminish clarity. As Claiborne (1986, p. 38) puts it, "Redundancy—using more words than you need to make your meaning clear—is like using too much water when you brew coffee; it dilutes the flavor, and discourages people from consuming it." Use as few words and syllables as possible to make important points with clarity. Write in simple English and omit jargon. In most cases, proceed from the general to the specific.

3. *Good writing is grammatical.* Although the English language is constantly evolving, there are established conventions that facilitate communication, and good writers follow them. Rules of word usage and sentence structure are taught in writing courses and emphasized in widely available books on English usage. Conventions of special relevance to the preparation of journal articles are emphasized in the *Publication Manual of the American Psychological Association* (APA, 1994). Certain errors are especially common when students first attempt scientific writing. Table 9.2 lists 25 of them.

Like oral presentation skills, writing skills are acquired. Once the basic rules of English are mastered, polish is attained by repeatedly writing manuscripts, securing feedback on them, and preparing revisions until they are accepted by a critical audience. Over time, good writers learn to be critical readers of their own writing; they can judge whether their writing is clear or not and make necessary modifications.

Although skill in using the language is necessary for writing a journal article, it is not sufficient. To merit publication, an article must be clear, but

TABLE 9.2. Twenty-Five Common Writing Errors

1. Using *data* as a singular noun.
2. Using ambiguous terms (e.g., "this," "that") without a clear referent.
3. Using sexist language.
4. Confusing *i.e.* (*id est*, "that is") and *e.g.* (*exempli gratia*, "for example").
5. Failing to indicate page numbers for quotes.
6. Failing to use parallel form.
7. Changing verb tense without reason.
8. Confusing the meaning of *affect* and *effect*.
9. Misplacing modifiers.
10. Failing to ensure the accuracy of references.
11. Beginning a sentence with a number that is not written out, a lowercase abbreviation, or a symbol.
12. Failing to hyphenate compound modifiers.
13. Using *which* when *that* is appropriate.
14. Confusing the appropriate use of colons and semicolons.
15. Using *since* as a synonym for *because*, which it is not.
16. Failing to ensure agreement of subject and verb.
17. Failing to use active voice.
18. Using the pronoun *who* to refer to nonhumans.
19. Using *consequate*, which is not a word.
20. Using split infinitives unknowingly.
21. Using *is comprised of* when *is composed of* is appropriate (the whole comprises the parts, and is *composed of* them).
22. Confusing the meaning of *can* and *may*.
23. Using double negatives.
24. Referring to other species as "infrahumans," "subhumans," or "animals," not as "nonhumans" or "other animals," when comparing them with us.
25. Using the ampersand when referring in text to references with two or more authors.

it must also be appropriately organized and significant in content. The following section describes the appropriate contents of the various sections of research articles and offers suggestions for writing those sections.

Parts of a Journal Article

Most research articles are organized into six major sections: abstract, introduction, methods, results, discussion, and references. Also characteristically included as part of a submitted manuscript are a title page, footnotes, tables, figure captions, and figures. Most journal submissions begin with the title page, then proceed sequentially through the abstract, introduction, methods, results, discussion, references, footnotes, tables, figure captions, and figures.

Despite this ordering, we recommend that authors begin by writing the methods section, then proceed in order to the results, introduction, and discussion. Compile the reference section as the other major sections are written, and complete the minor sections at the end of the project. The rationale for writing an article in this sequence is that the methods section is likely to be nearly complete even before the study is finished. The reason is that experimental methods must be clearly spelled out to secure permission from review boards for the use of human or nonhuman subjects and to enable experimenters to conduct the study. One must analyze the results of a study before making a decision to disseminate them, so this section is not hard to write. Completing the methods and results sections provides a critical mass of words; an author who has come that far is likely to press on. Moreover, the results of a study dictate, in part, how the study should be introduced (e.g., the aspects of the data that will be emphasized determine which prior articles are most relevant to the study being written up). The discussion section considers issues raised in the introduction; thus, it is reasonable to write the introduction and have it at hand and fresh in mind while writing the discussion.

Although some authors allow months or years to pass after completing a study before they begin writing a manuscript describing its results, this practice cannot be recommended. As time passes, enthusiasm often wanes. People disappear, memory fades, records are misplaced. The end result, too often, is that the planned paper is never written. To prevent this, one or more participants in a study should begin writing a manuscript as soon as it becomes clear that the study has yielded noteworthy results.

The best source of general information about preparing journal articles is the *APA Publication Manual*. The summaries of the various sections of an article provided below are consistent with, but far less detailed than, the information contained therein.

1. *Title page*. This page lists the title of the article, the authors and their affiliations, and a suggested running head. The title should be relatively short (roughly 50–75 characters) and informative. The running head, which usually appears at the top of every other page in the published article, should be very short (fewer than 50 characters). Some journals use a blind review format (i.e., referees aren't told whose work they are evaluating) and therefore require a second title page on which authors or affiliations do not appear.

2. *Abstract*. The abstract is a brief (about 150-word) summary of the contents of an article. It should describe the purpose, methods, and findings of the study, and the author's conclusions concerning those findings. References usually are not cited in the abstract. It is important that the abstract provide an accurate and informative capsule summary of the article, because this summary will be published in *Psychological Abstracts*, a

widely used reference source. To aid in indexing, some journals require that authors provide a list of key words. These lists often, but not always, appear after the abstract.

3. *Introduction.* The introduction sets the study in context. It (1) summarizes the general topic of interest, (2) describes the experimental question and its importance, and (3) proves a rationale for the procedures used to explore that question. Characteristically, the introduction begins with a historical overview of research and theorizing relevant to the current study. Coverage should be summary and selective, not detailed and comprehensive. It is reasonable to assume that readers are generally familiar with the literature and to refer them to other sources (e.g., articles and chapters that provide reviews of the literature) for further information.

Once a background is established, the specific question of interest in the current study is introduced. It is critical to demonstrate the importance of that question and to make clear how your study relates to previous findings and conclusions. Insofar as possible, justify the purpose and procedures of the study in a general way. Although it is appropriate to point out shortcomings in previous studies, avoid unnecessarily offending colleagues. When results of previous studies are inconclusive or when there are theoretical controversies, make sure that coverage is fair and unbiased.

4. *Methods.* The methods section describes how a study actually was conducted. By convention, this section usually contains *subjects, apparatus,* and *procedures* subsections, although other divisions may be used as appropriate. The fundamental rule in preparing the methods section is to provide sufficient detail to allow readers to replicate the study.

The subjects subsection provides information about the number of participants in the study and their important characteristics. With human subjects, it is convention to describe how they were selected and persuaded to participate, their general demographic characteristics (i.e., age, sex, geographic location, institutional affiliation), and any special characteristics that might affect their sensitivity to the intervention. With nonhuman subjects, most authors specify species, sex, age, experimental history, deprivation status (if relevant), and housing conditions.

The apparatus subsection describes any apparatus or materials used in a study. Readily available equipment usually is identified by model number and manufacturer; unusual apparatus is described specifically. Metric units are used in this section and throughout the manuscript.

The procedures subsection details each step in the experiment. It describes the independent variable and the manner in which specific subjects were exposed to it (i.e., the experimental design). This subsection also describes control conditions, steps taken to ensure the integrity of the independent variable (see Chapter 3), and exceptions to the usual experimental arrangement (i.e., any special treatment accorded particular subjects). A good procedures section describes who did what to whom and specifically how and why they did it.

5. *Results.* This section summarizes the data collected in a study. The usual strategy is to first describe the author's conclusions concerning important findings, then to justify these conclusions by presenting and discussing actual data. Data can be reported in the text or in tables, and they can be analyzed graphically through the use of figures or statistically through the calculation of inferential measures. Chapters 7 and 8 describe strategies for presenting and analyzing data.

Before submitting an article, check all figures, tables, and inferential statistics for accuracy. If you use inferential statistics, be sure that they are appropriate for your data. If the tests that you use are not obviously appropriate, defend their use.

6. *Discussion.* This section interprets results by relating them to the experimental question, to previous findings, and to established concepts and issues. Basically, the discussion clarifies the applied or theoretical importance of the reported study.

A good discussion section contributes information beyond that contained in the introduction and results sections and does not substantially repeat material contained in those sections. It is reasonable to provide multiple interpretations of observed results if they are merited, but refrain from providing interpretations not supported by logic and data. It is poor form to build and destroy straw men.

Although candor dictates that conclusions be qualified when findings are equivocal or methods suspect, do not dwell on weaknesses. Seriously flawed work does not merit publication and cannot be improved through candor or suggestions for improvement. Suggestions for further research should be limited to the nonobvious or avoided entirely. Most often, the latter tack is preferable.

Digression ruins discussions. Pursue major points in a clear and logical order, making sure to provide informative transitions. Digress only for a sterling reason, and indicate where the digression begins and ends, and why it is included. Stay focused.

7. *References.* The references section lists all sources cited in the article, nothing more—and nothing less. The *APA Publication Manual* describes the appropriate format for listing many kinds of references. Table 9.3 shows the APA format for referencing journal articles, book chapters, books, and papers presented at conferences.

8. *Other sections.* Chapter 7 details the preparation of figures, figure captions, and tables. Footnotes are self-explanatory. It is convention to acknowledge the funding agency if research is grant-supported, and ap-

TABLE 9.3. Examples of APA Format for References

IN THE REFERENCES SECTION

Articles

Hall-Johnson, E., & Poling, A. (1984). Preference in pigeons given a choice between sequences of fixed-ratio schedules: Effects of ratio values and duration of food delivery. *Journal of the Experimental Analysis of Behavior, 42,* 127–135.

LeSage, M., Makhay, M., DeLeon, I., & Poling, A. (in press). The effects of *d*-amphetamine and diazepam on schedule-induced defecation in rats. *Pharmacology Biochemistry and Behavior.*

Books

Poling, A. (1986). *A primer of human behavioral pharmacology.* New York: Plenum Press.

Poling, A., Gadow, K., & Cleary, J. (1991). *Drug therapy for behavior disorders: An introduction.* New York: Pergamon Press.

Book chapters

Poling, A. (1992). Consequences of fraud. In D. J. Miller & M. Hersen (Eds.), *Research fraud in the behavioral and biomedical sciences* (pp. 140–157). New York: Wiley.

Poling, A., & Cross, J. (1993). State-dependent learning. In F. van Haaren (Ed.), *Methods in behavioral pharamcology* (pp. 245–256). Amsterdam: Elsevier.

Conference presentations

Poling, A. (1991, August). *The practical importance of basic research in behavioral pharmacology.* Invited address presented at the American Psychological Association Conference. San Francisco, California.

IN TEXT

All sources

If the author or authors are named in the narrative, cite the year of publication in parentheses:

As Wolf (1993) suggested ...

An early report by Baer, Wolf, and Risley (1968) ...

If the narrative refers to a work without mentioning the author(s), cite the author(s) and year of publication in parentheses. If multiple works are cited, list them alphabetically across authors and chronologically within authors:

Four studies have investigated the effects of A on B (Ableson, 1962; Smith, Klein, & Cofars, 1981, 1982) ...

If there are one to six authors, name all authors the first time the work is cited. Thereafter, cite only the first author's surname followed by "et al." Do not underline "et al.," and use a period only after "al.":

In the first issue of the *Journal of Applied Behavior Analysis,* Baer, Wolf, and Risley (1968) described seven important characteristics of applied research. The seven dimensions described by Baer et al. (1968) were termed technological, behavioral, applied, analytic, effective, conceptual systems, and generality.

propriate to acknowledge all persons who contributed substantially to a project.

Once a manuscript is prepared, have all coauthors check it carefully for content, clarity, format, and organization. Spelling and grammatical errors make for hard reading and bias referees against recommending publication. Avoid them.

Summary and Conclusions

A study can benefit the culture that pays for it only when its results reach members of that culture. Data that are not disseminated cannot serve as a goad to action, whether by scientists or practitioners. Therefore, there is considerable pressure for scientists to share the results of their work. This chapter has briefly described some of the ways in which researchers communicate and has offered practical suggestions for researchers interested in making conference presentations or publishing articles.

The tentative rules offered to guide article preparation should prove useful for students who are preparing a master's thesis or a doctoral dissertation. Although these documents characteristically are larger than a journal article, primarily due to the inclusion of a more general literature review and (especially in the case of a dissertation) multiple experiments, they are written in much the same way.

References

Ainslie, G. W. (1974). Impulse control in pigeons. *Journal of the Experimental Analysis of Behavior, 21,* 485–489.

American Psychological Association (1982). *Ethical principles in the conduct of research with human participants.* Washington, DC: Author.

American Psychological Association (1985). *Guidelines for ethical conduct in the care and use of animals.* Washington, DC: Author.

American Psychological Association (1990). Ethical principles of psychologists (amended June 2, 1989). *American Psychologist, 45,* 390–395.

American Psychological Association (1994). *Publication manual of the American Psychological Association.* Washington, DC: Author.

Ary, D., Jacobs, L. C., & Razavieh, A. (1990). *Introduction to research in education.* Orlando, FL: Holt, Rinehart, & Winston.

Ayllon, T. (1963). Intensive treatment of psychotic behavior by stimulus satiation and food reinforcement. *Behavior Research and Therapy, 2,* 87–97.

Ayllon, T., & Azrin, N. (1968). *The token economy: A motivational system for therapy and rehabilitation.* New York: Appleton-Century-Crofts.

Ayllon, T., & Haughton, E. (1964). Modification of symptomatic verbal behavior of mental patients. *Behavior Research and Therapy, 2,* 87–97.

Ayllon, T., & Michael, J. (1959). The psychiatric nurse as a behavioral engineer. *Journal of the Experimental Analysis of Behavior, 2,* 323–334.

Baer, D. M. (1977). Perhaps it would be better not to know everything. *Journal of Applied Behavior Analysis, 10,* 167–172.

Baer, D. M., Wolf, M. M., & Risley, T. R. (1968). Some current dimensions of applied behavior analysis. *Journal of Applied Behavior Analysis, 1,* 91–97.

Baer, D. M., Wolf, M. M., & Risley, T. R. (1987). Some still-current dimensions of applied behavior analysis. *Journal of Applied Behavior Analysis, 20,* 313–327.

Ballard, K. D. (1986). Group designs, within-subject designs, case-study designs, and qualitative methodologies in educational and psychological research. *New Zealand Journal of Educational Studies, 21,* 42–54.

Barber, T. X. (1976). *Pitfalls in human research.* New York: Pergamon Press.

Barlow, D. H., & Hayes, S. C. (1979). Alternating-treatments design: One strategy for comparing the effects of two treatments in a single subject. *Journal of Applied Behavior Analysis, 12,* 199–210.

Barlow, D. H., & Hersen, M. (1984) *Single case experimental designs: Strategies for studying behavior change.* Elmsford, NY: Pergamon Press.

Barlow, D. M., Hayes, S. C., & Nelson, R. O. (1983). *The scientist-practitioner: Research and accountability in clinical and educational settings.* New York: Pergamon Press.

Bates, R. P., & Hanson, H. B. (1983). Behavioral assessment. In J. L. Matson & S. E. Breuning (Eds.), *Assessing the mentally retarded* (pp. 27–64). Orlando, FL: Grune & Stratton.

Bellack, A. S., & Hersen, M. (1988). *Behavioral assessment.* New York: Pergamon Press.

Bijou, S. W. (1955). A systematic approach to an experimental analysis of young children. *Child Development, 26,* 161–168.

Bijou, S. W. (1957). Patterns of reinforcement and resistance to extinction in young children. *Child Development, 28,* 47–54.

Bordens, K. S., & Abbott, B. B. (1991). *Research design and methods: A process approach.* Mountain View, CA: Mayfield.

Bowen, R. W. (1992). *Graph it!: How to make, read, and interpret graphs.* Englewood Cliffs, NJ: Prentice-Hall.

Braam, S., & Poling, A. (1983). Development of intraverbal behavior in mentally retarded individuals through transfer of stimulus control procedures: Classification of verbal responses. *Applied Research in Mental Retardation, 4,*279 –302.

Campbell, D. T., & Stanley, J. C. (1963). *Experimental and quasi-experimental designs for research.* Boston: Houghton-Mifflin.

Campbell, D. T., & Stanley, J. C. (1966). *Experimental and quasi-experimental designs for research.* Chicago: Rand McNally.

Catania, A. C. (1992). *Learning.* Englewood Cliffs, NJ: Prentice-Hall.

Christensen, L. B. (1994). *Experimental methodology.* Needham Heights, MA: Allyn & Bacon.

Cohen, J. (1960). A coefficient of agreement for nominal scales. *Educational and Psychological Measurement, 20,* 37–46.

Cohen, J. (1977). *Statistical power analysis for the behavioral sciences.* New York: Academic Press.

Cohen, J. (1990). Things I have learned (so far). *American Psychologist, 45,* 1304–1312.

Cohen, J. (1994). The Earth is round ($p < .05$). *American Psychologist, 49,* 997–1003.

Cohen-Mansfield, M. J., Marx, M. S., & Werner, P. (1992). Observational data on time use and behavior problems in the nursing home. *Journal of Applied Gerontology, 11,* 111–121.

Cooper, J. O., Heron, T. E., & Heward, W. L. (1987). *Applied behavior analysis.* New York: Macmillan.

Cozby, P. C. (1993). *Methods in behavioral research.* Mountain View, CA: Mayfield.

Critchfield, T. C. (1993). Preparing a successful convention presentation: A listener's perspective. *Experimental Analysis of Human Behavior Bulletin, 11,* 4–6.

Dachman, R. S., Alessi, G. J., Vrazo, J., Fuqua, R. W., & Kerr, R. H. (1986). Development and evaluation of an infant-care training program with first-time fathers. *Journal of Applied Behavior Analysis, 19,* 221–230.

Davison, M., & McCarthy, D. (1988). *The matching law: A research review.* Hillsdale, NJ: Erlbaum Associates.

Dawes, R. M. (1988). *Rational choice in an uncertain world.* San Diego: Harcourt Brace Jovanovich.

De Luca, R. V., & Holborn, S. W. (1992). Effects of a variable-ratio reinforcement schedule with changing criteria on exercise in obese and nonobese boys. *Journal of Applied Behavior Analysis, 25,* 671–679.

De Prospero, A., & Cohen, S. (1979). Inconsistent visual analysis of intra-subject data. *Journal of Applied Behavior Analysis, 12,* 573–579.

de Villiers, P. (1977). Choice in concurrent schedules and a quantitative formulation of the law of effect. In W. K. Honig & J. E. R. Staddon (Eds.), *Handbook of operant behavior* pp. 233–287). Englewood Cliffs, NJ: Prentice-Hall.

Dews, P. B. (1955). Studies on behavior. I. Differential sensitivity to pentobarbital of pecking performance in pigeons depending on the schedule of reward. *Journal of Pharmacology and Experimental Therapeutics, 113,* 393–401.

Dickinson, A. M., & Gillette, K. L. (1993). A comparison of the effects of two individual monetary incentive systems on productivity: Piece rate pay versus base pay plus incentives. *Journal of Organizational Behavior Management, 14,* 3–82.

Dillman, D. A. (1978). *Mail and telephone surveys: The total design method.* New York: John Wiley.

Domjan, M., O'Vary, D., & Greene, P. (1988). Conditioning of appetitive and consummatory sexual behavior in male Japanese quail. *Journal of the Experimental Analysis of Behavior, 50,* 505–519.

Dunlap, G., & Koegel, R. L. (1980). Motivating autistic children through stimulus variation. *Journal of Applied Behavior Analysis, 13,* 101–117.

Elliot, S. N., & Von Brock Treuting, M. (1990). The Behavior Intervention Rating Scale: Development and validation of a pretreatment acceptability and effectiveness measure. *Journal of School Psychology, 28,* 43–51.

Falk, J. L. (1967). Control of schedule-induced polydipsia: Type, size, and spacing of meals. *Journal of the Experimental Analysis of Behavior, 10,* 199–206.

Falk, J. L. (1971). The nature and determinants of adjunctive behavior. *Physiology and Behavior, 6,* 577–588.

Ferster, C. B., & Skinner, B. F. (1957). *Schedules of reinforcement.* New York: Appleton-Century-Crofts.

Fisher, R. A. (1925). *Statistical methods for research workers.* Edinburgh: Oliver & Boyd.

Forehand, R. L., & McMahon, R. J. (1981). *Helping the noncompliant child: A clinician's guide to parent training.* New York: Guilford Press.

Foxx, R. M., & Rubinoff, A. (1979). Behavioral treatment of caffeinism: Reducing excessive coffee drinking. *Journal of Applied Behavior Analysis, 12,* 335–344.

Frisch, C., & Dickinson, A. M. (1990). Work productivity as a function of the percentage of incentives to base pay. *Journal of Organizational Behavior Management, 11,* 13–33.

Fuqua, R. W., & Schwade, J. (1986). Social validation of applied behavioral research: A selective review and critique. In A. Poling & R. W. Fuqua (Eds.), *Research methods in applied behavior analysis: Issues and advances* (pp. 85–98). New York: Plenum Press.

Gadow, K. D., & Poling, A. (1988). *Pharmacotherapy and mental retardation.* Boston: College-Hill Press.

Gajar, A., Schloss, P. J., Schloss, C. N., & Thompson, C. K. (1984). Effects of feedback and self-monitoring on head trauma youths' conversation skills. *Journal of Applied Behavior Analysis, 17,* 353–358.

Geller, E. S. (1991). Where's the validity in social validity? *Journal of Applied Behavior Analysis, 24,* 179–184.

Gilgen, A. R. (1982). *American psychology since World War II: A profile of the discipline.* Westport, CT: Greenwood Press.

Gottman, J. M., & Glass, G. V. (1978). Analysis of interrupted time-series experiments. In T. R. Kratochwill (Ed.), *Single-subject research: Strategies for evaluating change* (pp. 197–235). New York: Academic Press.

Gould, S. J. (1981). *The mismeasure of man.* New York: Norton.

Gravetter, F. J., & Wallnau, L. B. (1990). *Statistics for the behavioral sciences.* St. Paul, MN: West.

Green, L., Fisher, E. B., Jr., Perlow, S., & Sherman, L. (1981). Preference reversal and self-control: Choice as a function of amount and delay. *Behavior Analysis Letters, 1,* 43–51.

Harris, S. L., Handleman, J. S., & Alessandri, M. (1990). Teaching youths with autism to offer assistance. *Journal of Applied Behavior Analysis, 23,* 297–305.

Hartmann, D. P. (1984). Assessment strategies. In D. H. Barlow & M. Hersen (Eds.), *Single case experimental designs* (pp. 107–139). New York: Pergamon Press.

Hartmann, D. P., & Hall, R. V. (1976). The changing-criterion design. *Journal of Applied Behavior Analysis, 9,* 527–532.

Hartmann, D. P., & Wood, D. D. (1982). Observational methods. In A. S. Bellack, M. Hersen, & A. E. Kazdin (Eds.), *International handbook of behavior modification and therapy* (pp. 234–277). New York: Plenum Press.

Hawkins, R. P. (1982). Developing a behavior code. In D. P. Hartmann (Ed.), *Using observers to study behavior: New directions for methodology of social and behavioral science* (pp. 21–35). San Francisco: Jossey-Bass.

Hawkins, R. P., & Dobes, R. W. (1977). Behavioral definitions in applied behavior analysis: Explicit or implicit. In B. C. Etzel, J. M. LeBlanc, & D. M. Baer (Eds.), *New directions in behavioral research: Theory, methods, and applications. In honor of Sidney W. Bijou* (pp. 167–188). Hillsdale, NJ: Erlbaum Associates.

Hawkins, R. P., & Dotson, V. A. (1975). Reliability scores that delude: An Alice in Wonderland trip through misleading characteristics of interobserver agreement scores in interval recording. In E. Ramp & G. Semb (Eds.), *Behavior analysis: Areas of research and application* (pp. 359–376). Englewood Cliffs, NJ: Prentice-Hall.

Hawkins, R. P., & Fabry, B. D. (1979). Applied behavior analysis and interobserver reliability: A commentary on two articles by Birkimer and Brown. *Journal of Applied Behavior Analysis, 12,* 545–552.

Hayes, S. C., Rincover, A., & Solnick, J. V. (1980). The technical drift in applied behavior analysis. *Journal of Applied Behavior Analysis, 13,* 275–286.

Henry, G. O., & Redmon, W. K. (1990). The effects of performance feedback on the implementation of a statistical process control (SPC) program. *Journal of Organizational Behavior Management, 1,* 23–46.

Herrnstein, R. J. (1961). Relative and absolute strength of response as a function of frequency of reinforcement. *Journal of the Experimental Analysis of Behavior, 4,* 267–272.

Hersen, M., & Barlow, D. H. (1976). *Single case experimental designs: Strategies for studying behavior change.* New York: Pergamon Press.

Heyduke, R. G., & Fenigstein, A. (1984). Influential works and authors in psychology: A survey of eminent psychologists. *American Psychologist, 39,* 556–559.

Horner, R. D., & Baer, D. M. (1978). Multiple-probe technique: A variation of the multiple baseline. *Journal of Applied Behavior Analysis, 11,* 189–196.

Howell, D. C. (1992). *Statistical methods for psychology.* Boston: PWS-Kent.

Huitema, B. E. (1986a). Autocorrelation in behavioral research: Wherefore art thou? In A. Poling & R. W. Fuqua (Eds.), *Research methods in applied behavior analysis: Issues and advances* (pp. 187–208). New York: Plenum Press.

Huitema, B. E. (1986b). Statistical analysis and single-subject designs: Some misunderstandings? In A. Poling & R. W. Fuqua (Eds.), *Research methods in applied behavior analysis: Issues and advances* (pp. 209–232). New York: Plenum Press.

Huitema, B. E. (1988). Autocorrelation: 10 years of confusion. *Behavioral Assessment, 10,* 253–294.

Hull, C. L. (1952). *A behavior system.* New Haven: Yale University Press.

Isaac, S., & Michael, W. B. (1981). *Handbook in research and evaluation, 2nd ed.* San Diego: EDITS Publishers.

Iverson, I. H., & Lattal, K. A. (Eds.) (1991a). *Experimental analysis of behavior: Part 1.* New York: Elsevier.

Iverson, I. H., & Lattal, K. A. (Eds.) (1991b). *Experimental analysis of behavior: Part 2.* New York: Elsevier.

Iwata, B. A., Pace, G. M., Kalsher, M. J., Cowdery, G. E., & Cataldo, M. F. (1990). Experimental analysis and extinction of self-injurious escape behavior. *Journal of Applied Behavior Analysis, 23,* 11–27.

Jarema, K., LeSage, M., & Poling, A. (1995). Schedule-induced defecation: A demonstration in pigeons exposed to fixed-time schedules of food delivery. *Physiology and Behavior, 58,* 195–198.

Johnson, S. M., & Bolstad, O. D. (1973). Methodological issues in naturalistic observation: Some problems and solutions for field research. In L. A. Hammerlynck, L. D. Handy, & E. J. Mash (Eds.), *Behavior analysis: Areas of research and application* (pp. 7–67). Champaign, IL: Research Press.

Johnston, J. M., & Pennypacker, H. S. (1986). The nature and functions of experimental questions. In A. Poling & R. W. Fuqua (Eds.), *Research methods in applied behavior analysis: Issues and advances* (pp. 55–83). New York: Plenum Press.

Johnston, J. M., & Pennypacker, H. S. (1993a). *Readings for strategies and tactics of scientific research.* Hillsdale, NJ: Erlbaum Associates.

Johnston, J. M., & Pennypacker, H. S. (1993b). *Strategies and tactics of scientific research.* Hillsdale, NJ: Erlbaum.

Jones, C. A., LeSage, M., Sundby, S., & Poling, A. (1995). Effects of cocaine in pigeons responding under a progressive-ratio schedule of food delivery. *Pharmacology, Biochemistry and Behavior, 50,* 527–531.

Kazdin, A. E. (1977). Assessing the clinical or applied significance of behavior change through social validation. *Behavior Modification, 1,* 427–453.

Kazdin, A. E. (1980). Acceptability of alternative treatments for deviant child behavior. *Journal of Applied Behavior Analysis, 13,* 259–273.

Kazdin, A. E. (1982). *Single-case research designs.* New York: Oxford.

Kazdin, A. E. (1992). *Research design in clinical psychology.* Boston: Allyn & Bacon.

Kazdin, A. E., & Kopel, S. A. (1975). On resolving ambiguities of the multiple-baseline design: Problems and recommendations. *Behavior Therapy, 9,* 912–922.

Kennedy, C. H. (1994). Manipulating antecedent conditions to alter the stimulus control of problem behavior. *Journal of Applied Behavior Analysis, 27,* 161–170.

Keppel, G. (1982). *Design and analysis: A researcher's handbook.* Englewood Cliffs, NJ: Prentice-Hall.

Koegel, R. L., & Egel, A. L. (1979). Motivating autistic children. *Journal of Abnormal Psychology, 88,* 418–426.

Kratochwill, T. R., & Wetzel, R. J. (1977). Observer agreement, credibility, and judgement: Some considerations in presenting observer agreement data. *Journal of Applied Behavior Analysis, 10,* 133–139.

Kuhn, T. S. (1970). *The structure of scientific revolutions.* Chicago: University of Chicago Press.

Kupfersmid, J. (1988). Improving what is published: A model in search of an editor. *American Psychologist, 43,* 635–642.

Lagomarcino, A., Reid, D. H., Ivancic, M. T., & Faw, G. D. (1984). Leisure-dance instruction for severely and profoundly retarded persons: Teaching an intermediate community-living skill. *Journal of Applied Behavior Analysis, 17,* 71–84.

Lattal, K. A., & Gleeson, S. (1990). Response acquisition with delayed reinforcement. *Journal of Experimental Psychology: Animal Behavior Processes, 16,* 27–39.

Lavelle, J. M., Hovell, M. F., West, M. P., & Wahlgren, D. R. (1992). Promoting law enforcement for child protection: A community analysis. *Journal of Applied Behavior Analysis, 25,* 885–892.

LeSage, M., Makhay, M., DeLeon, I., & Poling, A. (1994). The effects of *d*-amphetamine and diazepam on schedule-induced defecation in rats. *Pharmacology, Biochemistry and Behavior, 48,* 787–790.

Lindsey, O. R. (1956). Operant conditioning methods applied to research in chronic schizophrenia. *Psychiatric Research Reports, 5*, 118–139.

Lykken, D. (1970). Statistical significance in psychological research. In D. Morrison & R. Henkel (Eds.), *The significance test controversy* (pp. 267–279). Chicago: Aldine.

Maisto, S. A., Galizio, M., & Conners, G. J. (1995). *Drug use and abuse.* New York: Harcourt.

Makhay, M., Alling, K., & Poling, A. (1994). Effects of cocaine on fixed-ratio responding of rats: Modulation by required response force. *Pharmacology, Biochemistry and Behavior, 48*, 511–514.

Marholin, D., Touchette, P. E., & Stewart, R. M. (1979). Withdrawal of chronic chlorpromazine medication: An experimental analysis. *Journal of Applied Behavior Analysis, 12*, 150–171.

Mawhinney, T. C., & Gowan, G. R., III. (1991). Gainsharing and the law of effect as the matching law: A theoretical framework. *Journal of Organizational Behavior Management, 11*, 61–75.

Mazur, J. E. (1991). Choice. In I. H. Iverson & K. A. Lattal (Eds.), *Experimental analysis of behavior: Part 1* (pp. 219–250). New York: Elsevier.

McEwan, J. (1995). Graphic analysis of quantitative data: Towards a new methodology. Unpublished doctoral dissertation, University of Waikato, Hamilton, NZ.

Medawar, P. B. (1984). *The limits of science.* New York: Harper & Row.

Michael, J. (1974). Statistical inference for individual organism research: Mixed blessing or curse? *Journal of Applied Behavior Analysis, 7*, 647–653.

Michael, J. (1980). Flight from behavior analysis. *Behavior Analyst, 3*, 1–24.

Millar, A., & Navarick, D. J. (1984). Self-control and choice in humans: Effects of video game playing as a positive reinforcer. *Learning and Motivation, 15*, 203–218.

Miller, D. L., & Kelly, M. (1994). The use of goal setting and contingency contracting for improving children's homework performance. *Journal of Applied Behavior Analysis, 27*, 73–84.

Moore, J. (1990). A special section commemorating the 30th anniversary of *Tactics of scientific research: Evaluating experimental data in psychology* by Murray Sidman. *Behavior Analyst, 13*, 159–161.

Morrison, D. E., & Henkel, R. E. (1970). *The significance test controversy.* Chicago: Aldine.

Mosk, M. D., & Bucher, B. (1984). Prompting and stimulus shaping procedures for teaching visual-motor skills to retarded children. *Journal of Applied Behavior Analysis, 17*, 23–34.

National Institutes of Health (1986). *Public Health Service policy on humane care and use of laboratory animals.* Bethesda, MD: Department of Health and Human Services.

Navarick, D. J. (1982). Negative reinforcement and choice in humans. *Learning and Motivation, 15*, 203–218.

O'Donnell, J. M. (1985). *The origins of behaviorism: American psychology, 1920–1970.* New York: New York University Press.

Page, T. J., & Iwata, B. A. (1986). Interobserver agreement: History, theory, and current methods. In A. Poling & R. W. Fuqua (Eds.), *Research methods in applied behavior analysis: Issues and advances* (pp. 99–126). New York: Plenum Press.

Parsonson, B. S., & Baer, D. M. (1986). The graphic analysis of data. In A. Poling & R. W. Fuqua (Eds.), *Research methods in applied behavior analysis: Issues and advances* (pp. 157–186). New York: Plenum Press.

Perone, M. (1991). Experimental design in the analysis of free-operant behavior. In I. Iverson & K. A. Lattal (Eds.), *Experimental analysis of behavior: Part 1* (pp. 135–171). Amsterdam: Elsevier.

Peterson, L., Homer, A. L., & Wonderlich, S. A. (1982). The integrity of independent variables in behavior analysis. *Journal of Applied Behavior Analysis, 15*, 477–492.

Picker, M., & Poling, A. (1982). Choice as a dependent measure in autoshaping: Sensitivity to frequency and duration of food delivery. *Journal of the Experimental Analysis of Behavior, 37,* 393–406.

Picker, M., & Poling, A. (1984). Effects of anticonvulsants on learning: Performance of pigeons under a repeated acquisition procedure when exposed to phenobarbital, clonazepam, valproic acid, ethosuximide, and phenytoin. *Journal of Pharmacology and Experimental Therapeutics, 230,* 307–316.

Poche, C., Brouwer, R., & Swearington, M. (1981). Teaching self-protection to young children. *Journal of Applied Behavior Analysis, 14,* 169–176.

Poling, A. (1979). The ubiquity of the cumulative record: A quote from Skinner and a frequency count. *Journal of the Experimental Analysis of Behavior, 31,* 136.

Poling, A. (1985). Reporting interobserver agreement: Another difference in applied and basic behavioral psychology. *Experimental Analysis of Human Behavior Bulletin, 1,* 5–6.

Poling, A. (1986). *A primer of human behavioral pharmacology.* New York: Plenum Press.

Poling, A., & Foster, M. (1993). The matching law and organizational behavior management revisited. *Journal of Organizational Behavior Management, 14,* 83–97.

Poling, A., & LeSage, M. (1995). Evaluating psychotropic drugs in people with mental retardation: Where are the social validity data? *American Journal on Mental Retardation* (in press).

Poling, A., & Picker, M. (1987). Behavioral effects of anticonvulsant drugs. In T. Thompson, P. Dews, & J. Barrett (Eds.), *Neurobehavioral pharmacology* (pp. 157–192). Hillsdale, NJ: Erlbaum Associates.

Poling, A., & Thompson, T. (1977). Suppression of ethanol-maintained lever pressing by delaying food availability. *Journal of the Experimental Analysis of Behavior, 28,* 271–283.

Poling, A., Gadow, K. D., & Cleary, J. (1991). *Drug therapy for behavior disorders: An introduction.* New York: Pergamon Press.

Poling, A., Kesselring, J., Sewell, R. G., & Clear, J. (1983). Lethality of pentazocine and tripelennamine combinations in mice housed individually and in groups. *Pharmacology, Biochemistry and Behavior, 18,* 103–105.

Poling, A., Monaghan, M., & Cleary, J. (1980). The use of human observers in psychopharmacological research. *Pharmacology, Biochemistry and Behavior, 13,* 243–246.

Poling, A., Picker, M., Grossett, D., Hall-Johnson, E., & Holbrook, M. (1981). The schism between experimental and applied behavior analysis: Is it real and who cares? *Behavior Analyst, 4,* 93–102.

Poling, A., Schlinger, H., Starin, S., & Blakely, E. (1990). *Psychology: A behavioral overview.* New York: Plenum Press.

Poling, A., Smith, J., & Braatz, D. (1993). Data sets in organizational behavior management: Do we measure enough? *Journal of Organizational Behavior Management, 14,* 99–116.

Rachlin, H., & Green, L. (1972). Commitment, choice, and self-control *Journal of the Experimental Analysis of Behavior, 17,* 15–22.

Ragotzy, S. P., Blakely, E., & Poling, A. (1988). Self-control in mentally retarded adolescents: Choice as a function of amount and delay of reinforcement. *Journal of the Experimental Analysis of Behavior, 49,* 191–199.

Reaves, C. C. (1992). *Quantitative research for the behavioral sciences.* New York: John Wiley.

Redmon, W. K., & Lockwood, K. (1987). The matching law and organizational behavior. *Journal of Organizational Behavior Management, 8,* 57–72.

Renfrey, G., Schlinger, H., Jakubow, J., & Poling, A. (1989). Effects of phenytoin and phenobarbital on schedule-controlled responding and seizure activity in the amygdala-kindled rat. *Journal of Pharmacology and Experimental Therapeutics, 248,* 967–973.

Repp, A. C., Deitz, D. E., Boles, S. M., Deitz, S. M., & Repp, C. F. (1976). Differences among common methods for calculating interobserver agreement. *Journal of Applied Behavior Analysis, 9,* 109–113.

Risley, T. R. (1970). Behavior modification: An experimental-therapeutic endeavor. In L. A. Hamerlynck, P. O. Davidson, & L. E. Acker (Eds.), *Behavior modification and ideal mental health services* (pp. 103–127). Calgary, Alberta, Canada: University of Calgary Press.

Rosenfarb, I., & Hayes, S. C. (1984). Social standard setting: The Achilles heel of informational accounts of therapeutic change. *Behavior Therapy, 15,* 515–528.

Rosnow, R. L., & Rosenthal, R. (1989a). Definition and interpretation of interaction effects. *Psychological Bulletin, 105,* 143–146.

Rosnow, R. L., & Rosenthal, R. (1989b). Statistical procedures and the justification of knowledge in psychological science. *American Psychologist, 44,* 1276–1284.

Ross, D. M., & Ross, S. A. (1982). *Hyperactivity: Research, theory, and action.* New York: John Wiley.

Salvia, J., & Ysseldyke, J. E. (1981). *Assessment in special and remedial education.* Boston: Houghton-Mifflin.

Sedlmeier, P., & Gigerenzer, G. (1989). Do studies of statistical power have an impact on the power of studies? *Psychological Bulletin, 105,* 309–316.

Sidman, M. (1960). *Tactics of scientific research: Evaluating experimental data in psychology.* New York: Basic Books.

Skinner, B. F. (1938). *The behavior of organisms.* New York: Appleton-Century-Crofts.

Skinner, B. F. (1953). *Science and human behavior.* New York: Macmillan.

Skinner, B. F. (1956). A case study in scientific methods. *American Psychologist, 11,* 221–233.

Skinner, B. F. (1966). Operant behavior. In W. K. Honig (Ed.), *Operant behavior: Areas of research and application* (pp. 12–32). New York: Appleton-Century-Crofts.

Skinner, B. F. (1974). *About behaviorism.* New York: Knopf.

Skinner, B. F. (1979). *The shaping of a behaviorist.* New York: Knopf.

Solnick, J. V., Kannenberg, C. H., Eckerman, D. A., & Waller, M. B. (1980). An experimental analysis of impulsivity and impulse control in humans. *Learning and Motivation, 11,* 61–77.

Tawney, J., & Gast, D. (1984). *Single subject research in special education.* Columbus, OH: Charles E. Merrill.

Thompson, D. M. (1978). Stimulus control and drug effects. In D. E. Blackman & D. J. Sanger (Eds.), *Contemporary research in behavioral pharmacology* (pp. 159–237). New York: Plenum Press.

Tudor, R. M., & Bostow, D. E. (1991). Computer-programmed instruction: The relation of required interaction to practical application. *Journal of Applied Behavior Analysis, 24,* 361–368.

Tufte, E. R. (1983). *The visual display of quantitative information.* Cheshire, CT: Graphic Press.

Tukey, J. W. (1991). The philosophy of multiple comparisons. *Statistical Science, 6,* 100–116.

Turner, J. T. (1991). Participative management: Determining employee readiness. *Administration and Policy in Mental Health, 18,* 333–341.

Ullmann, L. P., & Krasner, L. (Eds.) (1965). *Case studies in behavior modification.* New York: Holt, Rinehart, & Winston.

Ulman, J. D., & Sulzer-Azaroff, B. (1975). Multielement baseline design in educational research. In E. Ramp & G. Semb (Eds.), *Behavior analysis: Areas of research and application* (pp. 359–376). Englewood Cliffs, NJ: Prentice-Hall.

Ulrich, R., Stachnik, T., & Mabry, J. (Eds.) (1966). *Control of human behavior.* Glenview, IL: Scott, Foresman.

U.S. Department of Health and Human Services (1993). *Preliminary estimates from the 1992 National Household Survey on Drug Abuse: Selected excerpts.* Washington, DC: Substance Abuse and Mental Health Services Administration.

Van Houten, R. (1979). Social validation: The evolution of standards of competency for target behaviors. *Journal of Applied Behavior Analysis, 12,* 581–591.

Werle, M. A., Murphy, T. B., & Budd, K. S. (1993). Treating chronic food refusal in young children: Home-based parent training. *Journal of Applied Behavior Analysis, 26,* 421–433.

Whalen, C. K., & Henker, B. (1986). Group designs in applied psychopharmacology. In K. D. Gadow & A. Poling (Eds.), *Methodological issues in human psychopharmacology* (pp. 137–222). Greenwich, CT: JAI Press.

Wilkenfield, J., Nickel, M., Blakely, E., & Poling, A. (1992). Acquisition of lever-press responding with delayed reinforcement: A comparison of three procedures. *Journal of the Experimental Analysis of Behavior, 58,* 431–443.

Winer, B. J. (1971). *Statistical principles in experimental design.* New York: McGraw-Hill.

Witt, J. C., & Elliot, S. N. (1985). Acceptability of classroom intervention strategies. In T. R. Kratochwill (Ed.). *Advances in school psychology,* Vol. 6 (pp. 252–288). Hillsdale, NJ: Erlbaum Associates.

Wittkopp, C., Rowan, J., & Poling, A. (1991). Use of a feedback procedure to reduce set-up time in a manufacturing setting. *Journal of Organizational Behavior Management, 11,* 7–22.

Wolf, M. M. (1978). Social validity: The case for subjective measurement or how applied behavior analysis is finding its heart. *Journal of Applied Behavior Analysis, 11,* 203–215.

Zuriff, G. E. (1985). *Behaviorism: A conceptual reconstruction.* New York: Columbia University Press.

Index

219